Methods and Perspectives in Geography

Methods and Perspectives in Geography

J. Beaujeu-Garnier
Professor of Geography, University of Paris

Translated by Jennifer Bray
Lecturer in Geography, University of Hong Kong

Longman
London and New York

Longman Group Limited London

Associated companies, branches and representatives throughout the world

Published in the United States of America by Longman Inc., New York

This translation first published 1976

Library of Congress Cataloging in Publication Data

Beaujeu-Garnier, Jacqueline.
Methods and perspectives in geography.

Translation of La géographie: méthodes et perspectives.
Bibliography: p.
Includes index.
1. Geography – Methodology. I. Title.
G70.B413 910'.1'8 75-23484
ISBN 0 582 48069 8

Set in IBM Baskerville 9 on 10pt
and printed in Great Britain by
Lowe & Brydone (Printers) Ltd, Thetford, Norfolk

Contents

Preface *vii*

Chapter one – What is geography? *1*

Geography reconsidered *1*
The unity of geography *4*
Physical geography is indispensable to human geography *9*
Physical geography is inseparable from human geography *14*
Two extreme examples of the role of the geographer *16*
Specific characteristics of geography *18*

Chapter two – Geographical method: review and reappraisal *23*

General processes *23*
Observation *26*
Explanation and models *30*
Precision – towards a new geography? *35*
Geographical method and scientific method *42*

Chapter three – Geographical space *44*

The geographer's space *44*
- A complex space *44*
- A concrete space *45*
- A coherent space *48*
- A variable and changing space *50*

Geographical space and economic space *52*
Static space: the concept of homogeneity *60*
The analysis of space *63*
- Identification of an element *63*
- Comparison of elements *67*
- Grouping of elements *69*
- Static elements and functional linkages *72*
- The static and the functional in geographical space *73*

The myth of the region 79
What is a region? 80
Types of region 81
Regions and regionalisation 86
Demythologising the word region 88

Chapter four – The division of space 89

Physiographic division 91
The influence of human activity 97
A visual division 99
Administrative divisions and the legacy of the past 100
Division according to dominant relationships 102

Chapter five – Conclusion – An outline typology for the division of space 104

Bibliography 107

Index 117

Preface

Although these discussions have deliberately focused on methodological issues, I am only too conscious of all that should have been written and to which this short volume does no more than allude. It is almost impossible in such a limited space to examine the need for revised approach to geography, to suggest a more systematic and comprehensive framework for spatial analysis, to consider fundamental spatial characteristics that have frequently been ignored or misunderstood, and finally to indicate new directions for research and publication in geography. However, lengthy discussions in recent years with both colleagues and students in France and overseas have convinced me that this is a useful and even essential task.

To write a definitive work on these topics at the present time would be pretentious, yet there is a danger of superficiality in so cursory a review. No book has ever demanded so much of my time, however, or the collection of so much bibliographic material. My sole aim has been to stimulate reflection and criticism and, I hope, to enlist the support of others who are plagued by similar doubts about our discipline.

Groups of research students in France are already working along the lines suggested. In collaboration with specialists from various other sciences and with representatives from both Government and independent organisations, they are preparing a series of monographs on the numerous questions raised in this preliminary survey, which merely summarises current research and new analytical techniques. It is not designed to incite revolution and destruction but perseverance in seeking a deeper understanding of the subject. Far be it from me to denounce the traditional geography of so many distinguished French and foreign scholars. New trends are in the air, however, and it would be as foolish to ignore them as to abandon all previously held views. We must try to synthesise tradition and innovation, and I for one am convinced that this is possible and that geography stands only to gain from the attempt.

Paris, 1967; Zermatt, 1971

I

What is geography?

It may seem paradoxical to begin these reflections on the nature of geography with a question that casts doubt on its very existence, or at least on its definition. For surely there is both a long tradition of geography and numerous publications in the subject? There have been geographers for more than 2000 years, and for at least a century even in the modern sense of the word. Surely the term 'geographical' is currently accepted as denoting a quite specific methodology – in other words, is there really little point in discussing the issue?

Certainly not. One need only read some of the articles by well-known geographers to see that their points of view differ enormously, or note how the word 'geographical' is used by economists and political scientists, to realise that this usage often bears little relationship either to the aim of the discipline or to the complexity of its methodology. (The observations by Reynaud, 1970, are a useful introduction to this discussion.) Even more significantly, the diversity of teaching methods and subjects that constitute geography courses in many colleges should convince us of the essential urgency of the issue.

It is a question which must be faced at a time when many of the younger social and economic sciences are broadening their own horizons and becoming increasingly aware of the prestige that stems from their research. They find themselves in competition over their field of study, a broad field where numerous approaches are permissible but one in which the study of human societies is the common denominator. It is also a time when geographers themselves are recklessly and often unjustly questioning the nature, the unity and the objectives of their own subject.

Geography reconsidered

Geography has been criticised in turn for its lack of a sense of direction and imperialist aims, for its degree of fragmentation and for expansionism. Its supporters claim that it is a science of synthesis, drawing on factual information from numerous other disciplines: its detractors accuse it of being no more than a superficial jack-of-all-trades. Geography is proud of its concern with the direct observation of real-world situations, but this has often led to the contention that expression in terms of mathematical laws is impossible. There has been inevitable criticism of its descriptions, often considered too literary, of the infinite shades of meaning in its phraseology, and of an emphasis on the unique in its monographs.

Many specialists from neighbouring sciences have stated their own views on the objectives of geography: they have sometimes assigned it a very limited role as a system of classification, as a convenient means of fixing or estimating distances, or even as a routine descriptive inventory, the basis for more high-flown theorising. Geography and cartography have often been confused. But opinion continues to be seriously divided and disputes are rife, both in France and in western and socialist countries where there are active departments of geography. Is geography one subject or several? What is its precise scope? Others go so far as to ask – what is its future? Is it merely a vague shroud, one phase of diffuse knowledge which must inevitably disintegrate with the progress of science?

Many research workers do in fact pursue their own specialisms so far that they come to deny the existence of geography as a single subject. It is as though a doctor, on the grounds that he is a cardiologist, will examine only a patient's heart, no longer acknowledging the need for an understanding of the pathology of the whole human body. Sorre (1967) commented that 'far too many adjectives have been attached to the one discipline', and that this has resulted in a failure to recognise the true core of geography.

Certain geomorphologists, hypnotised by their laboratories, would gladly ignore the presence of man, while there are climatologists who seem more concerned with the operation of general meteorological laws in the atmosphere than with the consequences of climatic phenomena for specific environments at the Earth's surface. Conversely, many zealous human geographers would be happy to break away from physical geography, some aspects of which appear as little more than a formal spatial framework, or as a collection of irrelevant factors scarcely influencing even primitive human groups, and rendered quite superfluous by modern technology – a symbolic theatrical décor rather than an integral part of the drama of human societies.

Such attitudes would lead inevitably to the demise of geography and, more seriously, would destroy all possibility of studying and interpreting the complexity of location on the Earth's surface, which is the primary aim of the subject.

In numerous overseas countries, this dissociation already exists and has been semi-officially recognised. In the United States, for example, there is an extreme degree of specialisation. Research in human geography is often associated with the work and publications of sociologists and economists, while geomorphology is closely allied to geology, rarely touching on aspects of human occupancy and limited to the differentiation of physiographic regions, clearly-defined areas within which man does not appear. Is the concept of integrated geography still understood in America? Although there have been outstanding monographs on one particular continent (such as that by Preston James on Latin America), or on an individual country (Trewartha on Japan), or on a single region of the United States, for many geographers the question is a legitimate one. Individual schools of thought are lively and active, but there is an excessive degree of fragmentation. I remember once travelling to the Mid-West with two geographers, one of whom was a specialist in animal husbandry, the other in arable farming: both were highly competent observers, yet they were reluctant to comment when the point at issue did not fall strictly within their own subject. To this fragmentation of geography should be added the controversy over methodology and even over concepts.

> In recent years passions have been aroused primarily by a debate between advocates of deductive logic, model building, and quantification and those who have faith that inductive field studies lead eventually to a comprehensive personal wisdom. It is difficult, to say the least, for someone who thinks that regional synthesis represents the highest expression of the geographic 'art' to enjoy cordial discussion with a colleague who condemns this tradition as being 'unscientific'. Spokesmen for these complementary positions view each other in a harsh and distorting light: the model builder becomes a 'mechanic' and the champion of regional synthesis is exposed as a 'mystic' (Mikesell, 1969).

This might well be a comment on the predicament now facing geography in France, as well as a description of the situation that our transatlantic neighbours have reached.

Many specialists also tend to maintain or develop closer links with related sciences than with other offshoots from the parent discipline. In the United States geography evolved from geology via physiography, which was concerned with the influence of the physical environment on man. The establishment of the first department of geography in 1903, at the University of Chicago, was intended to provide a bridge between geology and climatology on the one hand and history, political economy and biology on the other. This created a tradition of scholars who were conscious of their central position between the natural and social sciences, who had generally received a basic training in geology, and whose research interests were focused on the interaction between man and the natural environment. Two pronounced trends can still be traced in America as a result of these origins: the first is the preservation of the link between geomorphologists and geologists, through their work on the physiographic division of space; the second is the close link between cultural geography and other social sciences such as anthropology and sociology. A close parallel between these two trends is apparent: physiography sought to understand the processes by which natural landscapes had evolved, just as human geography explained the formation of cultural landscapes. Joint research with anthropologists has remained relatively limited, however, whereas there has been extensive collaboration with sociologists, some of whom also claim to be human ecologists. Geographers and sociologists concerned with urban problems, for example, have often worked in close cooperation and in several cases have also integrated their results for publication.

Among certain French and Soviet geographers, a marked change of direction has recently taken place within economic geography, that is, towards a common meeting ground with economics. Until the Second

World War this branch of geography was traditionally concerned with the description of land use, of types of production and of industrial establishments, and confined itself to the use of descriptive statistics on their classification and distribution. This has been partially replaced by a more theoretical approach, based on research into the analysis of structures and functional linkages. The initial impetus behind this change came from the writings of a small group of economists whose main interest lay in the analysis of spatial processes and structures: von Thünen, Weber, Lösch, Hoover, Isard and one geographer – Christaller. An English translation of von Thünen's book of 1826, *Der isolierte Staat in Beziehung auf Landwirtschaft und Nationalökonomie*, appeared in London in 1966. Location theories for the major economic activities have been outlined in numerous recent works and a new movement has been launched, illustrated not only by joint publications but also by the foundation of the Regional Science Association in 1956 by Isard. Economic geographers were initially regarded as junior members, but they soon challenged this attitude and have in fact played a leading role in an association designed from the outset as a means of communication between all scientists interested in location analysis. A new quarterly publication first appeared in 1969, *Geographical Analysis: an international journal of theoretical geography*, confirmation of the fact that quantitative economic geography was by that time considered sufficiently mature to stand on its own feet.

Provoked and reinforced by the demand for a classification based on a Marxist–Leninist approach, heated arguments have recently been taking place in the USSR between the natural sciences to which physical geography belongs and the social sciences which include human geography. The school in Leningrad was resolutely maintaining the dualism while opinion in Moscow was divided: one influential group, partly represented by Gerasimov, was directing research primarily towards issues in physical geography but met with strong criticism. The death of Stalin in 1953 and the period of de-Stalinisation which followed gave Soviet scholars a certain freedom, reflected in publication of the book by Anuchin (1960), a fervent plea for a unified geography. And in 1965 Saushkin wrote that 'geographers are starting to complain that specialists in one field no longer understand specialists in another. . . . Such disintegration of geography is tantamount to death. . . . We need a theory of the unity of geography' (Saushkin, 1965).

Which of us as a true geographer is not equally conscious of the uniqueness and the relevance of geography? Which of us would not feel concerned by the loss of the integrated approach that is essential to geographical interpretation? Division and fragmentation would be a mortal blow. It is time to confront this malaise.

The unity of geography

The meaning of the word geography, according to its etymology, is ambiguous: the 'study of the Earth' is open to numerous interpretations. Those whose work initiated the development of modern geography included as many men who had been trained as naturalists like Humboldt, or as Earth scientists in Germany and geologists in the United States, as they did historians such as the majority of French scholars at the turn of the century. This simple fact illustrates both the obvious duality within geography and also the danger of disintegration which threatens the subject by virtue of its dual inheritance.

As recently as 1885, the English historian E. A. Freeman wrote that he did not see how geography could be recognised as an independent study in the universities for 'on the one hand, a large section belongs to the historian, and on the other side, the geologist claims a large section of the field' (Freeman, 1961). Today there are specialists from neighbouring sciences who seem to assume that certain aspects of geography are simply an extension of their own research: a number of geographers, too, feel that they would best ensure the smooth progress of their work by allying themselves to those departments with which they are bound anyway to collaborate. There are therefore particularly strong links between geologists and geomorphologists, between sociologists, historians and social geographers, and between economists and economic geographers.

Geography thus finds itself in a trebly uncomfortable situation. The first ambiguity relates to the meaning of the word itself. The second difficulty arises from the diversity of interests of the pioneers of modern geography. Finally, the third danger – and this is perhaps the most serious – is attributable to the present highly varied and complex field of study.

The long treatise by Hartshorne, *The Nature of Geography*, published in the United States in 1939, drew the attention of geographers to most of the issues that have been raised in the first part of this book. His approach was both historical and encyclopaedic: he reviewed a very large number of works published up to that time by the major schools of geography in western Europe and the United States, and discussed fundamental questions that every geographer should ask during the course of his research – what are the essential characteristics of the discipline? What is its proper definition? What logical position does it hold among different branches of knowledge – geophysics, history, economics, sociology? What degree of contact with these other subjects should be encouraged? What are the objectives, the methodology and the claims of regionalisation? Is geography properly to be called a science? This long volume, although sometimes rather tedious in presentation, is a sound and impassioned plea for the uniqueness and the unity of geography.

We should not let ourselves become disheartened for something called geography, something quite indispensable, does already exist. There is an obvious need for concern with the interaction between man and his environment, between human groups and their natural surroundings. Who else but the geographer would investigate this? The naturalist? The sociologist, economist, agronomist or town-planner? One could list many more specialists in related sciences and examine the objectives of their research. Many would seem primarily interested either in certain aspects of human behaviour or in particular features of the physical environment, or occasionally in the limited relationships between one specific characteristic of the environment and its very restricted application – the agronomist for example – but none are willing to place the two on the same level, to devote equal attention to both and above all to the whole set of relationships which binds them.

For surely this is the proper objective of geography? No-one can seriously believe that it must be confined to a study of the bare planet, although consideration of the physical environment as a whole (and not merely one aspect of this environment such as relief or climate) certainly deserves recognition as a special subdivision of the subject, and future developments should be designed to further the study of its influence on human societies. Yet however valuable it may be, such an approach could never constitute the whole of geography. For wherever these environments are located they are accessible to man, penetrable if not already explored, and are transformed by his settlements: with each day they are conquered if not organised. Human societies populate the greater part of the globe and their presence is so clearly inscribed on its surface, their activities so closely interwoven with the environment even where this has been successfully tamed, that it is impossible to subdivide the integrated whole, to separate the planet from its inhabitants or man from his habitat. 'Study of the evolution of space content on the Earth's surface is the fundamental research frontier' (Ackerman, 1958).

This has been widely appreciated and there is no lack of support for cohesion in geography. Gribaudi listed a considerable number of disciples in a recent article (1965), but also demonstrated that their position is a defensive one and that the dualism between physical and economic geography is frequently an adoption from socialist countries, where the concept of unity is no longer generally accepted but is a question of dialectics. Certain modern developments, however, such as those in applied geography (or more accurately 'applicable' geography), with their precise and comprehensive analyses extending over the whole field of geography, 'are not an agent for the dissolution of our subject, but rather a factor for unity' (Meynier, 1969).

There are nevertheless certain human geographers who, while acknowledging the principle of cohesion, define geography in such a way that only the role of man is emphasised. According to Le Lannou, human geography is the 'science of human occupance', which may seem an unexceptional statement: a few pages further on, however, he adds in relation to physical geography – 'however absorbing it may be to solve the mysteries of nature's engineering, this exercise is of little use unless the spirit of human geography, of geography in effect, pursues it further' (Le Lannou, 1949).

Research should not be discouraged, however, since the advancement of science can only come about when detailed analyses are carried out with the highest possible degree of precision, to facilitate subsequent comparison and generalisation, or when specialists concentrate all their efforts on one particular problem and pursue it in greater depth than would be required for a general survey. A geomorphologist may thus investigate processes of deposition and suggest a chronology, whereas for a satisfactory general interpretation of the geographical complex, a knowledge of the composition and areal extent of a particular deposit would be sufficient: research into the precise conditions under which it had accumulated and whether it was Oligocene or Pliocene, Würm or Riss in origin, would be unnecessary. This personal and more detailed research is nevertheless of value for the general advancement of knowledge and is essential to the overall development of geography. A chronology which might appear superfluous in a monograph, but which had involved the collection of data on soil and climatic conditions, might prove invaluable in comparing the agricultural potential of certain areas. One must not reject *a priori* the pushing of even the most specialised research to its absolute limits. Apart from all the practical applications, the spirit of geography is not such as to stop half-way through an explanatory or correlative sequence. In carrying on, however, the geographer enters an interdisciplinary field, and the same research could often be undertaken by someone with a more specialised training, such as in this case a geologist or pedologist.

Drawing inspiration from his immensely varied field of study, the geographer frequently discovers his own source material through analysing the results of specialised research undertaken by others, which he then uses selectively in his own general interpretation. This is certainly an enrichment, but there is equally a risk of distraction and of confusion with neighbouring disciplines.

One must emphasise that the primary objective of geography, synthesis, 'where it is most fully itself' according to Vidal de la Blache, not only demands extensive knowledge but also complete mastery of that knowledge. It would be a mistake to leave interpretation and synthesis exclusively to geomorphologists, who would do at least as well with a training in geology, or to climatologists who had turned their backs on meteorology, or to the so-called human geographers who would simply emulate sociologists, demographers or economists, less rigorously and less competently. One should therefore mistrust too narrow a specialisation. 'The trend towards specialisation perhaps reaches its extreme expression in America, but it is also apparent in France. It may be dangerous or fruitful, according to the spirit or training of its practitioners.' T. W. Freeman (1961), author of the above quotation, concludes an otherwise critical paragraph on a reassuring note: 'an apparently simple theme, like an inquiry into electricity, may lead to a wide range of investigations' which are nevertheless of relevance. This is true only if those who undertake research into specific problems have previously received a sound general training, that is, if they have acquired the 'spirit of geography'.

A rather improbable illustration of the many branches of knowledge from which geography may have descended is provided in an article by Winkler (1970) on the possible classification of Earth sciences: he cites at least seventy different subjects!

This trend towards the disintegration of our discipline is accentuated by the apparent lack of concern among geographers that all research should have a common systematic methodology, taking into account both the general advancement of knowledge and contemporary ideas in scientific philosophy. It may seem surprising that Piaget, in his volumes on genetic epistemology (1950) and on logic and scientific knowledge (1968), does not even mention the word geography. However, as we have either failed to appreciate or misjudged current movements in the philosophy of science, it is only to be expected that we in turn should be ignored.

Fundamental modern methods have not been the object of research or even of systematic evaluation across the whole field of geography, but have crept into the subject piecemeal and indiscriminately.

> The climatologist thus draws much in the way of basic methodology from the closely related disciplines of atmospheric physics and physics. The biogeographer draws much from the soil scientist, the biologist, the chemist and so on. The historical geographer tends to look to history, the economic geographer may look to economics. Thus methodological separatism has grown in geography as each subdivision of the subject has matured and as the number of specialised subdivisions has increased (Harvey, 1969).

This then is the fundamental dilemma: if geographers continue in the same vein, geography as we know it will cease to exist, for one subject cannot both span such a wide range of themes (relief, towns, climate, industries) and also study each by a different method. The only link – and how fragile and tenuous – would be the indirect study of their interaction when, for an isolated moment, physical and human elements in the environment were reunited: even this may well not happen in studies primarily of physical geography.

Conversely, however varied the starting points, if a single methodology is rigorously thought out and applied it may give an underlying coherence to the whole. It is not necessarily the objective of a science which ensures its unity, but rather the nature of the approach to that objective. Take sodium chloride or common salt, for example: the geologist examines the characteristics of the salt deposits themselves, the chemist analyses its chemical composition and properties, and the mineralogist its elemental structure. How complex then is the study of man, over whom dozens of sciences are in contention! Yet if, for example, the geographer merely adopts demographic techniques for the study of a human society, he ceases to be a geographer: he must develop his own methodology. As Tricart observed, 'geomorphology is quite distinct from geology, to which it nevertheless owes a great deal, in that it can be integrated with all other aspects of the physical environment and has broad regional characteristics' (Tricart, 1968). One could cite many other examples.

A geographical approach has already been summarised as the investigation from a geographical viewpoint of linkages and interrelationships: man is not studied in himself and for himself, but in relation to his environment, to his social groupings, and to the outward signs of his activities. There is near-unanimity that this should be the perspective of a true geographer, yet very few possess a clear idea of an appropriate methodology that would allow them to pursue this unique approach. Even where such a methodology is apparent, it is intuitive rather than itself the object of a systematic examination. Books such as those by Hartshorne (rather too concise and with an epistemology that directly or indirectly reflected contemporary thinking in Germany, the most important at that time) and Anuchin (a plea for the unity of geography, directed primarily towards Soviet geographers), or the more recent work of Harvey (a far more scientific volume, which draws heavily on Anglo-Saxon and particularly on American publications, and which is now the most significant work in this area of research) – books like these are rare. In contrast, collections of essays such as *La Géographie Française au Milieu du XXe Siècle* (Chabot, 1957), *The Science of Geography* (National Academy of Sciences, Washington, 1965), or even *Frontiers in Geographical Teaching* (Chorley and Haggett, 1965), are evidence of the wide range of methodological viewpoints. They illustrate also the lack of interest shown by geographers in methodological issues: this may reflect either the nature of their research, which is usually directed towards very specific problems, or the broad field of geography itself, which dissipates their efforts, their methods of research and the relevance of their immediate conclusions. The result is a danger of disintegration into infinite subdivisions, and this is undoubtedly one of the main reasons why other scientists, and even more seriously certain geographers, are asking themselves in all conscience what is the substance of geography and even whether it exists!

Physical geography is indispensable to human geography

There is general agreement that it is extremely difficult, if not impossible, to study the activities of human societies without considering the framework within which they operate, that is, the environment. But what are the limits to this framework? On what basis should it form part of a study in human geography?

From the outset, the founders of modern geography sought to understand the interrelationships between human activity and the physical environment. Ritter was among the first to come to grips with this whole complex and to stress the interdependence of all phenomena on the Earth's surface. The implications of this concept soon led to the rise of a narrow determinism exemplified in the works of Ratzel, whose thinking is still reflected in much contemporary geographical writing. 'Man's whole life and all his activities are under the deterministic influence of the physical environment: one must first study the elements in this environment in order to understand the ways in which men have adapted themselves to it' (Lefèvre, 1965). One of the best-known determinists was Huntington, a geologist by training but later converted to climatology, who tried to demonstrate an almost total interdependence between climate and the nature of civilisations. Such a rigid viewpoint is dangerous: although a certain degree of interaction is undeniable, even at the highest level of technological sophistication, relationships between man and his environment are diverse and constantly changing. Certain societies may appear to triumph over physical constraints, but they can do so only at immense cost and through large-scale projects which then remain as witnesses to the original degree of dependence.

It is of course possible at colossal expense to establish oil towns in the desert of Kuwait or in the Sahara, or to cultivate sugar beet on the denuded chalk of Champagne with the aid of advanced farming techniques and large quantities of fertiliser. Yet although the dependence of man on his environment can be diminished or overcome it will always find some expression, if only in massive initial or recurrent financial investment.

No-one disputes that this interaction is more marked in particular fields: a whole group of rural activities, certain types of industry, the routing of many communication networks and thus the direction of circulatory flows, the settlement of successive human communities . . . all are very closely linked both in origin and development to the degree of attraction or repulsion of the physical environment. Many aspects of human activity and of societies themselves, however, are far removed from these constraints, which now operate only through a whole series of intermediate links that must be patiently re-assembled. In contrast to the theoretical space of economists, the geographer's space is concrete, infinitely varied and variable, never identical in two areas: each geographical study is therefore a 'unique study' (Cholley, 1948).

This implies that the concept of the environment simply as a framework is inadequate – as a backcloth before which the players go through their movements but without contact or harmony. It does not imply, on the other hand, that numerous subdivisions are desirable – a method which has long been customary in so-called descriptive regional geography, notably in publications of the French school during the first thirty years of this century, and which is still common in many theses – where in highly-polished but quite independent chapters, all those features characterising a small area are discussed in minute detail. This happens particularly in joint works, where each specialist presents the end product of his own inquiries. It is no more than the preparation of the raw material of geography, an indispensable but nevertheless incomplete task. Such works cannot be considered geographical unless they contain a substantial final section, in which all the elements of the analysis are drawn together in an integrated overview, the essence of a geographical perspective. This underlines a fundamental point: geography should not be a random juxtaposition of data or even an orderly dissection – it is above all a carefully-prepared synthesis.

Le Lannou discusses this very well in several sections of his *Géographie Humaine*. One might re-iterate his words on the need for an awareness of the primacy of man, who

> far from weakening interest in research into the natural environment, should direct, justify and actively promote it, since there are problems actually within geomorphology and climatology that have been created by colonisation of the landscape, by sharp contrasts between two types of land use, and by differences in the agricultural calendars of two neighbouring rural communities. It is not surprising that many issues in the natural sciences should have been raised and resolved by geographers who were convinced of the ultimate dominance of their discipline by man (Le Lannou, 1949).

One might also mention his example of the striking portraits of the western Alps drawn by Raoul Blanchard (1938–56). The natural environment comes to life beneath his emotive and imaginative pen to present a living framework for human endeavour, and the questions an observer might ask receive a response that is both scientific and immediately understood. Each unit is isolated, examined and re-assembled: life clings to the slopes where the varied bedrock has been subjected to repeated erosion; it exploits the gifts of humidity – forests, hydroelectric power, the sea; it flees the *ubacs*, the most hazardous areas of the high valleys, and concentrates where low passes allow easy communication and on the *adrets*, slopes favourable both for cultivation and for building; it then sets out for the snowfields of the summits, a new face of the mountain for twentieth-century man to explore.

The power of suggestion founded on a keen appreciation of all the component parts, on a personal and detailed study of all related fields, on exhaustive reading of the most specialised scientific publications and theses to have appeared on the subject, should be presented to the reader as 'more of an art than a science' (Birot, 1965, Préface: Gilbert, 1960).

This example from the Alps illustrates the kind of balance that must be preserved in any analysis of the interaction between man and his environment. In certain branches of geography, such as the study of population, agriculture or even transport, the importance of the physical background is obvious. If, however, in following new lines of investigation, the human geographer ventures into studies that are primarily economic, the environment may seem more remote and of less immediate relevance. What therefore should its role be in this 'discontinuous space', this 'relative space'? (George, 1968.) Here too the geographer must be consistent: the study of flows or the operation of industrial mechanisms must not lead to a neglect of the central issue – the spatial distribution of human enterprises and their ultimate repercussion on the natural environment. The geographer must confine his interest primarily to concrete space.

Whether a factory has been built on a substratum of chalk or on sands reclaimed from the sea and artificially hardened, as in the case of Usinor at Dunkirk, governs both the cost of the infrastructure and the technology of construction. Modern large-scale industries requiring extensive areas of flat land have often been deliberately sited on alluvial plains, which may be more or less well consolidated and which until now have been avoided by urban sprawl. These include the region to the southeast of Lyon, the banks of the Tyne–Tees estuaries in Northumberland–Durham and, on a much vaster scale, the reclaimed marshes along the wide bays of the Atlantic coast of North America and of Guanabara to the north of Rio de Janeiro. The conditions under which the raw material or energy requirements of an industry are extracted and transported are also closely related to the physical environment. The scars of open-cast mining and the disposal of waste products from processing plants have caused derelict landscapes in former coalmining areas of England and the United States. Mineral prospecting itself requires an intimate knowledge of geology, topography and hydrology. Specific industrial locations are obviously governed by physical factors, and few examples are as clear-cut as that of cotton manufacturing. During the Industrial Revolution, the availability of pure running water and the high level of humidity together assured the prosperity of Lancashire. The establishment of large textile factories in this area was indirectly stimulated at a later date both by various local human factors and also by the import of cotton from the old American

South where soil, temperature and humidity conditions favoured the development of plantations.

Industries are also subject to physical constraints through their demand for water. In the mid-nineteenth century, rapid expansion in the textile industries of Roubaix–Tourcoing could only take place after borings had revealed a deep aquifer of Carboniferous Limestone, able to yield a precious liquid that had been in increasingly short supply. Endless proposals could be devised for costly and technologically ingenious projects that would allow great industrial centres and large cities, with their huge concentrations of humanity, to procure the water that is essential to their survival. However, the problems of discharging used water and of the pollution of rivers and shorelines are no less urgent. The galloping pace of industrialisation in Japan has presented an immediate and alarming danger to the purity of water for the rice fields in certain areas, while the excessive pumping of groundwater has resulted in land subsidence, threatening the low-lying districts of a number of towns. Problems associated with the use of water as a means of transport include such familiar examples as the smelting of iron ore at coastal sites in both western Europe and the United States. This has already radically altered production costs, the degree of competition and consequently the general state of the market, and in future may affect even living standards in certain inland areas: the whole means of livelihood of the people of Lorraine, which because of its extensive ore deposits is a great interior iron-smelting region, is now seriously endangered.

It is not only particular types of manufacturing and the opening up of new markets that are influenced by specific physical factors: the texture of certain African soils has proved an obstacle to the use of conventional ploughs, while it was the exploitation of oil in the Sahara desert that forced Berliet to build the world's largest lorry.

Strong and diverse relationships, sometimes indirect but rarely insignificant, are therefore seen to link the physical environment and human activities. Many dubious and costly projects have been undertaken to no purpose through ignorance or denial of these links, for example, the building of dams in the Chusistan mountains of Iran, where both water supplies and plots of land were carefully allocated despite the absence of a local population to exploit the region! The engineers would have benefited considerably from the advice of geographers on this scheme.

There may be objections, however, to the assertion that the geographer need only study and explain the whole complex of physical factors in a summary or limited way, before claiming that his basic research into human problems is firmly grounded in an environmental framework. Recent research and theses, on the other hand, prove that many apparently incidental steps are invaluable and that specialised physical research may be highly relevant in an attempt to explain certain features of rural life, or even aspects of land capability.

While it may be true that all Frenchmen appreciate the quality of Burgundian wines, it was Rolande Gadille who first successfully tackled both 'the physical and human bases of this high quality viticulture' along the Côte Bourgignonne. Through laborious and detailed analyses of the micropedology and microclimatology of the area, she was able to show that the most favourable xerothermic conditions occur particularly on east-facing slopes, on the lower third of the hillside, and that

> the most productive vineyards are located where there is a critical proportion of clay in both the loams and fine gravels: this balance is affected also by the angle of slope of the talus and the depth of the soil profile. The characteristics of the Oligocene deposits on adjacent slopes coincide precisely with the long-observed relationships between wines produced from these slopes.

The calculation of a topo–pedological index enabled the author to outline optimum environmental conditions and this must now be tested against first-growth wines, by careful evaluation of 'the skill of the vine-grower and the science of the wine-maker on the one hand' and the quality of the vintage from vineyards of known high productivity on the other (Gadille, 1967).

The remarkable study of the agricultural life of the Morvan by Jacqueline Bonnamour is a perfect example of this type of integrated research. Certain agronomists have recommended the development of intensive cultivation, despite the fact that over 30 per cent of the area is dominated by forest and that elsewhere there is mixed land use, in which stockrearing and therefore pasture are extensive. But is this small over-exploited highland area in the Massif Central a condemned region? The answer lies within both physical and human geography. The author has attempted a dynamic study of pedogenetic factors, and has demonstrated that 'the evolution of soils is related to a highly complex equilibrium', an equilibrium so unstable that 'even where the slope does not exceed 15 per cent, the single turn of a ploughshare, the introduction of a new crop or even of a simple ground cover, is sufficient to alter the whole system'. But it takes a good farmer to make a good farm! In modifying the natural vegetation cover man affects the accumulation of humus: excessive resin-tapping, for example, accelerates the process of podsolisation and creates an unfavourable local microclimate. Foraging is now detrimental to 'those soils which require a careful rotation of arable and semi-intensive cultivation'. Thus Bonnamour, with a true geographical perspective, demonstrates through an initial pedological survey that what has condemned the Morvan is not the inherent quality of its soils, but rather their bad economic management, largely as a result of social factors such as isolation, ignorance, poverty and negligence (Bonnamour, 1965).

This clearly illustrates the vital and additional contribution of the geographer to the scrupulously-presented analyses of other specialists. Both in Burgundy and in the Morvan, agronomists, oenologists, geologists, botanists, economists, historians and many others have each concentrated on their own field – on soils, crops, methods of production, economic potential. Only a geographer can draw together the scattered threads and weave a fabric of the highest quality. Only a geographer can try to unravel the subtle relationships that exist between features of the landscape and resources exploited by man, by launching resolutely into hitherto unexplored depths of geographical complexity.

Physical geography is inseparable from human geography

Whereas human geographers may admit the necessity of considering certain elements in the physical framework, disputing only the importance of the actual details of analysis, there are many physical geographers who are quite prepared to eliminate human geography altogether. In so doing, they cease to be simply geographers and become specialists in various sub-branches, such as geomorphology or hydrology, which may be self-contained but which can no longer lay claim to the spirit of geography. However much they might wish it, it is sometimes quite impossible to ignore the presence of man.

There are in effect two types of link between physical factors and human activity that should be examined in specialist studies. The first, which has been alluded to above, may in specific circumstances be disregarded. A hydrologist often fails to consider the consequences for human communities of the physical features which he analyses in detail. It is perfectly feasible to describe terraces, and to trace the evolution of a slope or an erosion surface by analysing the deposits and the chronology of deposition, while remaining silent on the importance of these features for man. Although it is essential in determining land use, for example, to know the extent of the erosion of deposits on the low terrace of Alsace, or whether a particular area of East Anglia is covered by glacial drift, as well as the characteristics of individual drift deposits, the physical geographer may leave to others the task of evaluating their significance. In certain cases he arbitrarily curtails the sequence of causal linkages at a particular point, but in so doing he deviates from the true path of geography.

There exists a second link between physical factors and the presence of man, however, which makes their joint study absolutely essential. Man intervenes not only as a 'consumer' of the environment, but also as an agent in its transformation. The rapid spread of soil erosion in many countries is directly attributable to the removal of the vegetation cover. This is the well-known 'anthropic' or 'anthropogenetic' erosion (Tricart, 1953): the plantation of vast belts of forest across the Russian steppes was explicitly designed to bring about climatic change. The construction of barrages, such as that across the lake surrounding Brasilia, may modify both the local climate and the regional hydrology, besides transforming conditions of urban life. Similarly, certain of the grandiose schemes which are being devised in the Soviet Union would lead to a change in climatic conditions over vast regions and perhaps even in ocean temperatures. One cannot therefore neglect man if one wishes to preserve the study of the Earth in all its aspects, including obviously the most dynamic, rather than merely of spatial responses to the rhythm of geological time.

Many publications illustrate this close and essential interdependence. The distribution of dry valleys on the Russian steppes, for example, is primarily a function of climatic conditions, but once the traditional peasant communities destroyed the natural vegetation cover through over-cultivation, they released

> a new wave of pseudo-climatic anthropogenetic erosion, comparable in nature and intensity to the results of European

> colonisation in the American Mid-West: in losing its own characteristic features, the wooded steppe became indistinguishable from the ordinary steppe: ravines developed on the denuded slopes; wind deflation reached an alarming level on the interfluves (10 cm per year on an experimental plot near Kouibichev); sheets of running water ravaged soils which had already been disturbed by ploughing. The morphological extent of the true steppe has been extended by man (Tricart, 1953).

This article by Tricart, an eminent geomorphologist who has remained an outstanding geographer, is worth pursuing since it further emphasises the value of integrated analyses. Having examined in the first section how the interaction of natural and human factors led to the formation of devastating ravines, he then discusses the remedial measures taken by the Soviet government which, through human effort, will enable it to control the natural environment: the planting of forest belts around active gulley systems; careful management of the more gentle cultivated slopes; the introduction of complex systems of rotation, in which fodder crops 'which form a continuous dense cover and so hold the soil together' will occupy a considerable area; regular planting of forest screens over the vast areas exposed to wind erosion. Only three years after the start of this plan to transform the region, there was already a noticeable change in the local microclimate and ecological conditions.

The existence of vast expanses of alluvial silts along the river valleys of north-western Germany, covering the low terraces to a depth of several metres, has attracted the attention of many geomorphologists. Detailed research, and pollen analysis in particular, has enabled them to trace the deposition of these silts to the late Middle Ages, when clearance of the valleys and the development of cereal cultivation reached a level previously unknown.

> According to Müller-Wille, the proportion of arable land in the loess basin of Göttingen rose from 5 to 60 per cent between A.D. 400 and 1200. In the region of the Upper Weser, Jäger claims that the clearings of the late medieval period reduced the forest area by at least 25 per cent. This destruction of the forest and breaking up of the soil, with subsequent removal of the loess from vast areas through stream erosion, resulted in the extensive deposition of alluvium. The discharge of precipitation as surface run-off was a direct consequence of the clearance, and the massive accumulation of alluvial silts therefore coincides with the period of deforestation in the Middle Ages. It was essentially human interference with the natural vegetation that caused the sudden beginning of the widespread deposition of these silts (Mensching, 1951).

It would be difficult to illustrate more decisively the significance of human activity for dynamic geomorphology.

There are similar examples from many countries, but they are particularly spectacular in regions of sudden climatic change. A simple localised incident such as 'the construction of storm drains along a road-side may be a sufficient catalyst. Mass wasting, and in some years the wholesale transformation of a slope, is caused by an alteration to the run-off pattern and a subsequent increase in the rate of infiltration' (Millies-Lacroix, 1965). This is particularly characteristic of certain developing countries where western technology has been applied to the construction of a road network, but where preventive measures have not been taken. In other cases, a whole region may be condemned to destruction and sterility, including such well-known examples as the effect of landslides on the heavily deforested hill slopes of Brazil and the complete removal of fertile soils from the 'dust bowl' of the central American plains, when several years of intensive cereal cultivation were followed by a period of exceptional drought.

Human activity is therefore an important agent for change in the physical landscape. This activity may be random and spontaneous, revealed only *a posteriori*, but may also have been planned and organised, designed to produce carefully predetermined effects – the principle of Land Management. It is essential that geographers who are involved in studies of physical phenomena should be aware of this: not only will their own specialist research be improved and enriched, but they are less likely to draw partial conclusions that could be reached equally well by numerous other technologists. With a consistent methodology, they should be able to make far-reaching recommendations that can be applied by management consultants, who have a genuine need for studies of this kind. The uniqueness of a geographical approach is clearly recognised and appreciated in these essential and integrated surveys.

Two extreme examples of the role of the geographer

Having discussed the unity of the discipline, we must now consider where precisely the borderland of geography as a science lies. This debate has been continuing for many years and has always aroused passionate polemics. To throw some light on the issue if not to settle it – which would be presumptuous – I have chosen two examples, both extreme in my sense of the word, of the current activity of geographers. The first is the role of a young French geomorphologist, Pierre Rognon, in a team of petroleum scientists working in the Sahara; the second is the well-known research of an experimental geographer in the United States, William Applebaum, into the location of various types of commercial activity, and in particular the most rapidly-expanding form of retail distribution – supermarkets.

Pierre Rognon has himself defined the nature and importance of his own contribution.* In the early stages of the exploitation of an oil field, very general information on the stratigraphic succession and the extent of the permeable strata is sufficient, since the location of anticlinal structures by geophysical methods permits the rapid identification of promising

---

* This is a text drafted by him at my request.

drilling sites. Once these have been exhausted, however, recovery from small reservoirs related to stratigraphical traps requires the application of far more precise techniques. A geological structure is no longer considered as having a uniform lithology: the more porous sandstone bodies, which may contain local reserves of hydrocarbons, must be distinguished from the less well-defined deposits and impervious rocks. These sandstone bodies have an extremely variable form (lenses, channels, . . .) which must first be studied in comparable outcrops, often at a distance of several hundred kilometres from the sandstones themselves. This type of terrain analysis is feasible only in relatively flat areas, where the series remains homogeneous over great distances, and in arid regions where there are distinct outcrops. General laws must then be formulated to explain the distribution of the porous sandstones, in order to correlate the data with the potential oil reserves.

It is at this point that the geomorphologist intervenes, to explain both the origin of the sedimentary structures (the main bed of water-courses, moraines, offshore bars) and the spatial distribution of the sand-stones. Hundreds of thousands of core-samples, obtained from drilling during the intensive exploitation of the area, are available to help him relate his conclusions to the location of the petroleum deposits. Such an investigation entails close collaboration between the geomorphologist and the geologists, who are already familiar with the terrain from their studies of the outcrops, and with the specialists who have been drilling for sub-surface data. In the same way, unconformities which were assumed to be in a fixed plane during the initial exploitation are then studied in greater detail. For example, they may be a relict cuesta relief, with the more porous sands following the line of the wide consequent valleys (the unconformity of the Triassic 'red beds' on the Mississippian in Canada), or inselbergs with sandy peripheral cones (the palaeo-relief of quartzite fossilised during the Cambro-Ordovician in Kansas), or even glacial valleys with knob and kettle country (the central Sahara).

This type of research can only be carried out by a team of specialists, since it requires a knowledge both of the present physical land-scape and the lithology as revealed in the outcrops, and also of prospecting techniques for petroleum. The geomorphologist must deepen his own understanding of existing erosional and depositional features, and particu-larly of their spatial distribution, so as to detect the slightest evidence that would aid palaeographic reconstruction. Such research may also have a practical objective in parallel situations as in Canada, for example, where these palaeogeomorphological traps in general contain 10 per cent of the workable deposits of a mineral, rising to 30 per cent in southeast Saskatchewan where the palaeotopographies are particularly favourable.

From his initial geomorphological observations, therefore, the geographer together with other specialists may reconstruct concealed fossilised structures and, as in the above examples, contribute not only to the most profitable exploitation of mineral reserves but also to more reliable predictions relating to the construction of underground workings.

The work of William Applebaum and his team, obviously in a very different field, is significant in its demonstration that the rational location for each commercial enterprise (particularly a large plant requiring heavy initial investment) is a 'unique situation': a general economic model has proved inapplicable. There must be a direct inquiry into a whole range of factors, such as

> the objectives of the firm, its resources (in the field of finance, management), its policy of supply and operation, the degree of competition, market potential, demographic characteristics, consumer behaviour, environmental factors influencing overall economic trends, the transport system and legal regulations (zonation, taxes, monopolies); a sound strategy for the establishment of a supermarket also requires an intuitive sense of the right moment (Applebaum, 1968).

These lines from the preface to one of his two volumes are illustrated by numerous case studies. It is essential to analyse the relationships between a new consumer good (it is for the promoters to specify the attributes of this commodity) and the point at which it is to be introduced. The latter obviously includes examination of physical data such as the angle of slope and nature of the terrain, which may necessitate additional expenditure on infrastructure as proved recently by the careless or misinformed choice of certain large retailing sites in France. The synthesis, the final decision on the market potential and the actual point of impact, should be reserved for a joint team of promoters, geographers and economists.

Is there any connection between these two extreme examples? Why should geographers find themselves involved in both operations? Is it a chance occurrence or does it reflect the personality of the individual research worker, or even the nature of geography as a discipline? An answer to these questions would amount to a definition of geography. It is significant in both cases that features which are clearly inscribed on the Earth's surface form the basis for later analyses – morphological characteristics in the first example (nature of the relief and surface deposits, . . .) and human features in the second (settlements, means of transport, distances, population density, place of work and residence, . . .). This initial observation and description of the environment is the first stage in a search for explanatory mechanisms (processes of erosion and deposition, determination of income levels and consumer behaviour respectively), and for interpretations and conclusions that could both have useful application and also stand on their own as excellent geography: an understanding on the one hand of relict landscapes, of the former extent and erosion of surfaces that represent past morphological periods, and on the other of the features of a neighbourhood or district within a suburb or small town.

The significant point is surely that concrete spatial units are the starting point for all geographical research: the study of mechanisms then enables the geographer to trace the evolution of successive interrelationships, with the aim of reaching a comprehensive explanation.

Specific characteristics of geography

It is time to draw tentative conclusions from this discussion on the nature of geography and its undeniable uniqueness in relation to neighbouring disciplines, and thus to suggest a preliminary answer to the initial question – what is geography?

Firstly, on what is all research in geography based? Many authors have declared, although often in too simple or limited a sense, that it is 'the landscape'. I would rather define it as 'the observation of elements clearly inscribed on the Earth's surface'. These may form a natural landscape, with distinctive relief features or vegetation type, a rural landscape comprising both villages and farmland, an urban landscape, . . . and also the distribution network for a commercial enterprise or bank or a particular sphere of influence, both of which have visible indicators but which do not constitute a landscape in the classical sense of the term, that is, observable at first glance.

Although initial description of the chosen element is important, many other disciplines could claim that they also do this: a geologist observes rocks, a demographer or sociologist a specific aspect of human societies. What primarily determines the unique character of geography is that these elements are invariably complex. Rognon does not examine the infilling of a fossil valley for its own sake, but in relation to the surrounding environment, to the system of erosion which gave birth to it, and to the network of analogous valleys of which it forms part. This principle is even more convincing when applied not to a specialised branch of geography but to its most general aspects. The geographer cannot choose what his eyes encompass – houses, vegetation, rocks overhanging a valley: the whole complex must be his field of study.

The corollary of the above is the need to analyse relationships between quite distinct groups of facts. Contrary to the majority of other disciplines, research in geography cannot be contained within the limits of one field of study. The distribution of population, for example, cannot be explained by considering only demographic factors, or the economic prosperity of a valley merely through morphometric analysis.

Geography studies the relationships between highly diverse elements: this is again evident in integrated surveys, where the two basic components are the natural environment and settled human communities. The environment is the raw material, the object, complex both in its evolution and surface features and possessing its own internal dynamism, while man is the active force. This duality of object–instrument is particularly characteristic of geography as a discipline.

The symbiosis between the two is permanent. However ingenious the joiner or powerful his tools, he is not the servant of the plane or the mechanical saw, but only of his wood. Nor can one separate the sculptor or painter from his raw materials. The work of art is his point of contact: it expresses the quality and intensity of his relationship. In the same way, one cannot detach the natural environment from those who live there, who have adapted their way of life to its constraints or themselves transformed it. Geography is the study of their interaction. Because of this almost inevitable association of physical phenomena (and therefore Earth sciences) with human factors (social and economic sciences), geography is a 'crossroads discipline' and unlike many other sciences is not pre-occupied simply with facts, however complex, within a single field of study. The true geographer is both a man of science and a man of letters. Besides submitting to the discipline of the laboratory, he must be able to evoke through the richness of his vocabulary and the flexibility of his style the infinite variety of landscapes and of men. Gilbert, who in 1960 wrote that 'the art of describing a region . . . is quite as difficult as the art of describing the character of a human individual', has shown that many English novels in fact contain excellent descriptions of the geographical personality of a region.

On the other hand geography quite specifically aims to study various forms of interaction and, as Cholley so ably demonstrated, it seeks to determine the characteristics and reasons for these convergences. It isolates them: it considers them in space and time, but also requires a certain continuity and solidity and is therefore concerned only with relatively stable environments. The geographer is not a journalist: he does not ignore the present in so far as it marks the culmination of an evolutionary process, but he must not become pre-occupied with a single event in time.

What empirical procedures does the geographer adopt? In the past the starting point has always been the publication of a monograph, a broad general survey, which undoubtedly has a certain usefulness but which contributes only one stone to the building. Many stones are required – and a host of monographs – in order to reach the second stage, that of comparison followed by generalisation. General laws governing the evolution of karst landscapes could be formulated only when dozens of world-wide examples of karstic relief had been described: the preliminary study of thousands of towns allowed the major dimensions of 'urban geography' to be outlined. It was essential to study the demographic evolution of both urban and rural, primitive and developed human societies, and to design and publish hundreds of censuses, before something more than simply a series of regional monographs on population geography could be produced, or before any general human geography could be written. These are the necessary preliminary steps towards what is traditionally called 'general geography' or the study of comparative relationships, drawing parallels between unique facts in order to extract what they have in common. The initial monographs supply 'the basic raw material and the necessary comparative framework for hypothesis verification' (Baulig, 1959).

A further stage is to examine relationships on the 'horizontal' plane rather than between cause and effect, the 'vertical' plane. The German Varen (Varenius) was certainly the first to be concerned with general as opposed to special geography, but writing in the mid-seventeenth century he was ahead of his time. A great deal of analytical data is required before general rules can be derived. We can scarcely claim that such data is available today, which may partly explain the current flood of volumes on geomorphology, climatology, population geography, urban geography and transport geography. Many earlier attempts, while marking notable steps forward, have been shown to be premature. Among the events which signalled the maturity of our discipline, special mention must be made of the encyclopaedic work on physical geography by de Martonne, the first edition of which was published in 1909. This pioneer work drew together all that was known at that time and was so outstanding that it was translated into seventeen languages, assuring its author both international fame and an eminent role in world geography for more than a quarter of a century. However, more recent observations have either cast doubt or have expanded on many of the views held by this inspired scholar: several of his individual chapters, for example, now correspond to whole volumes in the only series on dynamic geomorphology currently published, under the joint editorship of Tricart and Cailleux. One can trace the expansion of the subject over a period of sixty years, while the list of works and authors consulted before presenting the final synthesis clearly indicates the breadth of the basic groundwork.

Following his initial observations, the geographer should always adopt the same procedures to complete his monograph or to produce a more comprehensive explanation and synthesis. He should analyse, describe and then attempt to answer the questions summarised by an English author at the turn of this century as 'Where? What? How? When?' He must draw on information from other sciences, extract from them what is necessary to his own analysis, and then outline relationships that are both original and often unexpected. He thus achieves this 'unique study', this reflection of a reality that is infinitely varied and variable: his primary aim is an exhaustive synthesis.

It is the delicate sifting of many specialist contributions that is the geographer's most crucial role: it is this which involves the greatest element of risk and which earns him the most virulent criticism. At any moment he may stray too far, tempted away from his own field by the science to which he has had recourse. The geomorphologist risks being engulfed by geology, mineralogy or chemistry, the urban geographer falling prey to sociology, urban technology or statistics. Yet it is of paramount importance not to lose sight of the intricate relationships between complex facts, the fine balance between multiple interrelationships. If this approach is not meticulously followed there will inevitably be criticism from workers in other disciplines. They will label as superficial and incompetent, often quite justifiably, anyone who pushes ahead without either having the requisite training or openly discussing his opinions with true specialists.

Precisely where, however, does research into interrelationships lead? Is it not a trap for our complacency or our inexperience? In the same context one might ask whether or not there are 'forbidden areas', facts which geographers must disclaim since they are outside their field of study? A pertinent example is that of the economic geographer Pred, who broadly studies 'the background to the growth of industries in American towns. He reached the conclusion that some of the distributions which he was studying were determined by the interplay of financial factors – and there he stopped, declaring that the study of these factors was the work of economists' (Claval, 1966).

Two extreme points of view are held by non-geographers. The first, defended by the anthropologist Gluckman in his book *Closed Systems and Open Minds* (1964), is that all investigations should be restricted to a clearly-defined field of study, that explanations should be sought within this field and not beyond it and interdisciplinary amateurism renounced. The second, put forward by Huxley in his contribution to *Man and his Future* (1963), is that there must be a unified approach to research with a common methodology and terminology. Neither of these attitudes is realistic. Interdisciplinary scientific research is absolutely essential, not only to the geographer. No fact relates only to one science: no science is monolithic or focused only on one attribute of one object. On the other hand, one must beware of a magma: progress and enrichment do not arise out of confusion. A sociologist, a geographer, an agronomist and an economist do not in fact have the same approach. If they make a joint study of a village, for example, or of rural life in a particular area, each specialist will contribute his own stone to the building: the same fact considered by each one of them will not take the same form, but provided that – and here we are in agreement – there is clear and open communication, their collaboration will be richer and more productive. Research into a common terminology is certainly necessary, but at the same time the uniqueness of the methodology appropriate to each science must be carefully preserved within collective research. The autonomy of geography must be asserted on condition that, as will become apparent in the following pages, its own methodology can be precisely defined. Professor Fenneman declared to a meeting of American geographers more than fifty years ago that 'if we are concerned about our independence, it is better to extend our field of study than to cling to the defence of its limits'.

This unashamedly aggressive point of view is tantamount to a profession of faith and is a valid conclusion to our assertion of the existence of geography, of its uniqueness and of its assured place among neighbouring disciplines. However, it would be dangerous to slide into complacency because of the underlying optimism in such a conclusion. Geography has a distinctive role, but we have seen just how vulnerable and hazardous this is because of the specific attributes of the subject. Instead of squandering the initiative on triumphant pyrotechnics we must firmly consolidate the foundations or risk going seriously astray.

2

Geographical method: review and reappraisal

General processes

A multilingual bibliography is of little help in the search for a coherent methodology. Without elaborating on the traditions of particular schools of geography, or on the personality of individual scholars (which must inevitably influence a discipline where the presentation of research findings is in a literary form), the logical sequence of reasoning is often difficult to identify. Beneath the apparent diversity, however, there are two fundamental methods of investigation: empirical-inductive and theoretical-deductive.

The inductive method requires a large number of case studies on a single theme, identical analyses of their analogous elements, comparison of results, generalisation, the formulation of a comprehensive explanation and ultimately of a theory. Brookfield (1964) suggested that the stages of inquiry in human geography could be either:

1. The establishment of generalisations by the study of interrelated phenomena over a wide range of territory
2. Detailed local inquiry to explain particular distributions
3. The organisation and classification of both generalised and local material so as to yield new and more explanatory generalisations

or the same procedures but in the order 2, 1, 3.

The deductive method involves preliminary exploration of all the theoretical ramifications of a problem and the construction of a model, that is, the formulation of a general proposition followed by empirical verification of the theory derived from this model. If the results are not satisfactory, the model must be modified and new relationships suggested – and so on.

It is conspicuous that this second method, while undisputed in the pure sciences, has only recently been applied in geography. There are a few outstanding examples of deductive research, however, which have undoubtedly been among the most productive of any discipline, to judge from the response they provoked among both geographers and other scientists. They include the normal cycle of erosion (W. M. Davis, 1899) and the central place theory of Christaller (1933). In both cases, a general theory has given rise to apparently inexhaustible debate, counterproposition and new hypotheses. Böventer (1969) suggested that Christaller's theory was the most original in the field of spatial economic analysis during the first fifty years of this century. Almost thirty years after publication of the original manuscript, Berry and Pred edited a bibliography devoted exclusively to central place study (1961).

Which of these two approaches should be followed in geographical research? Although one can agree with Harvey's recent statement (1969) that indiscriminate and alternate use of empirical-inductive and theoretical-deductive reasoning 'may be inevitable in the early stages of a discipline's development', it is surely time for geography to grow out of this period of prolonged adolescence.

At its simplest the inductive method seems the more likely to fit geographical reality, but it also reflects the personality of the author. However logical in its subsequent development it remains a personal methodology, and there is a danger that in the confused melting-pot of original ideas the overall plan will be submerged in a mass of detail or lost in uncontrollable complexity. With such a proliferation of initial case studies, it may be difficult to derive a general proposition from empirical observations, particularly where specialists are working independently: an element which appears unique may in fact be a chance occurrence, while n observations may coincide but the $(n + 1)^{th}$ destroys the whole framework. If an hypothesis is satisfactory, the outcome of a given series of analyses will be the grouping and classifying of elements, usually based on the straightforward comparison of external physical characteristics and other measurable parameters, and more rarely on data relating to structure since there is no predetermined frame of reference where a process of inductive reasoning is followed. Weber has already stated with respect to history that an isolated event is of no significance unless considered in relation to a regular pattern of events. The same reasoning can be applied to geography. 'The ultimate purpose of geographic analysis may be to understand individual cases', that is, 'formulating laws which can then be applied to explain particular instances' (Harvey, 1969).

The deductive method is based on a carefully and skilfully established framework. This should not lead to neglect of any essential characteristics of the element in question, but allows removal of irrelevant detail which threatens to mask significant correlations or hidden similarities. It encourages geographers to prune reality to fit a precast mould, but to strike the right balance between what can and cannot be omitted, or considered of secondary importance, is a delicate operation. Deductive reasoning should not result in the sacrifice of a wealth of careful observation, nor is it a substitute for the shrewd intuitive deduction of an observer, but it faces him with the discipline of a logical sequence of steps in his research. It has the enormous advantage over induction of being comparative – a research worker exploring a similar topic in any other country could immediately isolate the initial assumptions and common traits.

This approach is therefore a great deal more productive, as long as it is applied with flexibility and integrity. One must be ruthless in judging the extent to which the model twists reality, however, and whether or not such deviation can be accommodated in the final synthesis without altering the model itself, in which case fresh observations and hypotheses would be required. Perhaps it is acceptable for a subject like geography to combine the use of a sound deductive methodology with conceptualisation from a wide range of direct observations.

A major difficulty, which incites rebellion … geographers, lies in this degree of conceptualisatio… one begin to develop a general theory as the starting … reasoning? This should include the observation of co… awakening of a disciplined scientific curiosity, both of … into hypothesis formulation and the construction of a m… discussed, geography is unique in taking as its starting poi… tion of concrete facts. It is also quite clearly impossible for … that aims to be – and indeed must be – scientific, to contin… burdened with inspired but superfluous detail and an archaic …ical framework. Scientific theory is not simply 'a key to the puzzles of reality' but also has the power to predict (Bunge, 1966).

The geographer must try to explain diverse man–environment relationships over the whole globe, and participate in determining future developments through his contribution to resource management and planning. He must therefore pursue a rigorous scientific approach through comparative analysis and synthesis – or become someone who does no more than describe what he sees, who comments often with great skill, but who remains simply a figurative landscape painter and not the scholar he claims to be. Almost alone among specialists in any discipline, geographers are still presenting their research within a traditional framework, without considering epistemological issues and – particularly in France where these issues have scarcely been raised at all until the present time – they are increasingly alienating themselves from current scientific movements, and run the risk of 'expounding without reasonable foundation views which have generally been discredited in all other disciplines as well as in the philosophy of science' (Harvey, 1969).

These considerations might seem relevant only to 'generalisation' in geography (I have deliberately avoided the term 'general geography' since this could lead to confusion). There are in effect two geographical methods: a *descriptive* method – leading to the compilation of a monograph (on the population of Alsace, the industries of New England, the ice caps of Greenland, Lancashire, the Moscow region, . . .), each one describing an apparently unique feature – and a *comparative* method – a system for classification and explanation of spatially dispersed phenomena of the same order (cuesta relief, textile industries, urban regions, transcontinental railway networks, . . .).

This can only add to the confusion: the continued existence of two approaches is impossible if geography is to survive as a unified and coherent science. The uniqueness of a geographical fact is only an apparent uniqueness, resulting firstly from our own superficial understanding and secondly from simplistic reasoning, which makes any explanation in depth impossible. How can we explain, in the full sense of that word, what we cannot compare? Bunge (1966) highlighted the confusion between the *unique* and the *individual* case:

> Individual case implies generality, not uniqueness. For example, assume there is a theory that explains the existence of islands. There is only one Manhattan Island. Yet, if Manhattan Island conforms to the theory of islands, it is different

> ...other islands only in that the variables are in peculiar ...quantitative combination. Manhattan Island is an individual case, as are all other islands, and the theory is still applicable.

Kolotievskij (1967) has also discussed the antithesis between unique and common regional characteristics, and underlined the importance of the latter. If we do no more than examine the concept of uniqueness, we shall simply describe a sequence of events and obvious or subjective relationships, concentrating only on details of that uniqueness – details which are relevant as a means of identifying a particular unit but are not its essential characteristics. Suppose we toss a large number of assorted objects into water: the fact that some may float while others sink does not mean that each object, although apparently quite distinct, is not in fact obeying the same fundamental law. The experiment could be described according to either of the methods outlined above. The behaviour of each object in the water could be examined, but in terms of what – its external appearance, gross weight, shape, size, colour, origin, density, . . .? Much time and energy will be wasted in groping blindly for a common thread, which may never emerge at all. Alternatively, starting from a familiar and tested hypothesis, an experiment could be designed which allowed the behaviour of the objects to be explained immediately or, if necessary, predicted without actually conducting the test.

This obviously does not itself constitute geography. Although we can apply laws from the natural sciences to certain aspects of physical geography, those relating to man might seem less amenable. However, while the reactions and behaviour of an individual may be unpredictable, at least at our present level of understanding, this is not necessarily true of groups and societies. Sociologists and economists are already formulating general laws of evolution and causality and it is time that geographers followed their example, by suggesting and verifying parallel hypotheses. Certain economists consider that the geographer's only contribution is to provide observations from a series of case studies, in order to prove or disprove the theories which they themselves have developed (Perroux, 1954). While partly accepting this role, since our discipline claims to be concerned primarily with existing concrete features, we must also envisage collaboration at the stage of theory formulation.

It could justifiably be argued that a temporal or spatial occurrence is not *per se* geographical, and that there is no reason why a human 'occurrence' should be so. An individual initiative frequently sets in motion a variety of developments: a dynamic mayor, for example, can stimulate the process of urban expansion, but do the mechanics of this process follow clear and precise laws? This is what we must attempt to establish. W. M. Davis (1899), who can hardly be accused of modern opportunism, claimed that 'to exclude from geography the theoretical half of man's brain is like walking on one leg or seeing only through one eye'.

Observation

The monograph should not be rejected outright as something now irrelevant and archaic, but its scope and limitations must be appreciated.

In trying to produce a synthesis or a comparative analysis, we have all suffered from the disparate mass of data available and often have been forced to abandon the attempt. Take the analogy of two walls – the first is rapidly built of finely-bonded stones, no two alike but each in its correct position as part of an overall design: it is solid, visually satisfying, aesthetic as well as effective. The second wall is laboriously constructed from a jumble of large blocks and small stones, and is held together only by large quantities of mortar. It needs an outer layer of pebbledash to conceal the rough edges: if a single stone works loose the whole structure begins to crumble. Which of these two walls is the geographer hoping to build?

The monograph is relevant to geographical research in two ways. In the initial stages, the *exploratory* monograph allows us to develop new lines of approach and is crucial in suggesting new links and interrelationships. At this level, its inherent flexibility encourages a lively spirit of innovation, although the empirical methods themselves must be clearly defined. This basic raw material of geography must be of high quality and capable of being subjected to searching analysis, leading ultimately to the construction of a model and to theoretical propositions. The role of the *verificatory* monograph is to devise a series of tests which will then confirm or disprove a particular theory. It must therefore be far more rigorous in conception, and have as its primary objective the identification of unifying factors beneath an apparently diverse exterior.

There are two quite distinct stages in the preparation of a monograph: observation of an object and presentation of research findings. But what objects attract the geographer's attention? Although certain features can be reproduced in a laboratory, the geographer must continue to analyse components of the actual landscape: his workshop is the Earth's surface. Experimental work is already making an important contribution to research in physical geography, but is confronted both by the need to simplify reality into the framework of a model, and also by the problem of time. For example, when the disintegration of crystalline rocks is being studied under artificial conditions, through the alternate application of heat and cold, is it the treatment itself which is significant or its regular application over a limited or even over an unspecified time period? The laboratory is like a kitchen – the ingredients are added according to the recipe but the final dish is the result of blending them all together. There is therefore a risk of omitting an essential ingredient when trying to break down the total environment into a series of experiments, and this is particularly true of human and so-called regional phenomena.

We have so far considered only observation of the present landscape, but the computer is opening up new possibilities through the use of simulation models. These enable us to predict the necessary modifications to a given hypothesis if one or more variables are altered, and have great potential for the analysis of complex spatial problems. Whether in the field or in the laboratory, however, it is essential that geographers should make careful and precise observations at every stage of their research.

Scientific observation is not the simple or spontaneous activity one might have imagined. It must be guided by a desire for discovery yet follow strictly defined rules, and therefore demands qualities of both sharpness and precision. Even a simple observation, one that involves no

…nent but merely description within a predetermined framework, …lect the qualities (or inadequacies) of the observer as well as the …ibutes of a particular feature, since each one of us sees the world 'by …fraction through cultural and personal lenses of custom and fancy' (Lowenthal, 1961).

Despite the good intentions of an observer his description is inevitably biased. If two people simultaneously describe the same street, one of them a native of the town seeing it for the hundredth time and the other a new arrival, their accounts will show striking differences. The same would be true of two outside observers: their reports would reflect the size of their home town, the local climate (one from a cold and the other from a tropical area), their racial type and social background – or simply their individual sensitivity and, even more unpredictably, the mood of the moment. A single observer, viewing exactly the same landscape on two separate occasions, will also describe it in slightly different terms.

It is therefore hardly surprising that geographers who are content with this subjective impressionism can write colourful and convincing regional descriptions, and draw sketches that are bursting with life – only to declare themselves defeated by the uniqueness of geographical facts.

Both approaches are necessary: the direct impact – the steaming public baths, the smell of the earth, the deafening noise of machines, the jostling in the streets, the mud on the race track – and the greater objectivity of a judicious use of statistics. I have written 'use' rather than 'choice', for 'geography is a science in the sense that what facts we perceive must be examined, and perhaps measured, with care and accuracy. It is an art in that any presentation (let alone any perception) of those facts must be selective and so involve choice, and taste, and judgement' (Darby, 1962).

This careful observation usually enables the geographer to achieve two goals: first, the communicability of his observations, that is, the possibility of his research being understood and used by another specialist working on the same topic at the same or a later date. This is crucial to progress in any discipline and to the establishment of strict scientific procedures. Geographers are in a particularly vulnerable position since they borrow so much of their terminology from neighbouring sciences, often without appreciating its precise meaning – this is true especially with economic concepts. French geographers seem captivated by the magic of words and seduced by strange sounds, to which they attach a meaning all the richer as they differ from their French equivalents. 'Cuesta', a short Spanish word meaning little more than 'hillside' or 'slope', has become in French a distinctive relief type with its own characteristic features. Many foreign words pass from one language to another without translation, and in each case seem to acquire a new shade of meaning: much of the vocabulary of coastal and tropical geomorphology has been lifted from the English. This internationalisation of terms has obvious disadvantages. A further source of geographical terminology is the general adoption into the scientific literature of a purely local term denoting a morphological feature, including such words as aven, sotch and polje, as well as those now applied to a general relief type – karstic, appalachian, monadnock.

These quirks of modern international vocabulary are not peculiar to the geographer, but force him to exercise great care in recording his observations.

The second aim of scientific observation is to so describe the object or space under investigation that generalisations and comparisons can be made, leading to classification and a comprehensive interpretation. Progress towards explanation is accelerated as the number of observations increases. The precise analysis of even a few elements permits the formulation of an explanatory hypothesis or an embryonic theory, but many more are needed to verify or expand the theory, to modify or even ultimately to disprove it. Although a theory is often shown to be inadequate, or of only temporary value, it serves as a useful springboard for further research until an improved version is developed.

The evolution of Davisian theory is a good example: the normal cycle of erosion, as formulated by Davis in 1899, was primarily derived from his study of the magnificent stepped erosion surfaces of the Appalachians – and these horizontal tiers would certainly make a deep impression on any geographer seeing them for the first time. Davis's research was a remarkable contribution towards scientific geography since it comprised 'the stage by stage reconstruction of the evolutionary sequence, and the indication for each stage (as had not been attempted previously) of the interdependence of related land forms and of a series of idealised relief types – imaginative concepts but essentially derived by generalisation from observable reality' (Baulig, 1959). His theory therefore rested on two hypotheses: firstly, that normal erosion operated on a homogeneous highland massif, with long periods of tectonic and eustatic stability interrupted by sudden cataclysmic upheavals – leading to a world-wide adjustment of relief levels and a new cycle of erosion; secondly, that progressively senile valley forms corresponded to the increasing height of the erosion surfaces, in a cyclical relationship. He assumed a gradual and continuous process of levelling, with intermittent rejuvenation following localised incision, and a single erosive force acting on a stable and homogeneous, effectively passive, basement complex.

More recent observations have shown that the laws governing normal erosion differ from those envisaged by Davis and that other processes, such as sheet erosion in tropical and arid areas or periglacial frost shattering, are more likely to give rise to level surfaces. Continents are slowly shifting and warping, as well as being subject to tremendous unheavals. The variable resistance of the bedrock to erosion is also a significant factor – in other words, the concept of an unstable equilibrium between conflicting and variable forces should replace that of a steady linear development.

These later results might never have been obtained, however, had there been no initial challenge to stimulate other research workers into criticism of the theory (A. Penck, 1919; W. Penck, 1924; Tricart and Cailleux, 1950, 1957) or its defence (Baulig, Johnson, 1959). The evolution of the Limousin, for example, had been successively interpreted according to Davisian theory by Demangeon and Perpillou, both of whom used morphometric techniques. However, it was later completely reinterpreted by Beaujeu-Garnier and Bomer, who attached greater importance to detailed observation of the landscape, particularly of surface deposits and geological structures.

A new problem area in research, theoretical propositions and tentative explanations can all develop out of accurate observation. Many geographers adopt this approach intuitively and spontaneously and produce scientific geography without realising it. This is generally the case in France, whereas Anglo-Saxon geographers have given considerable thought to the processes involved. The French attitude, while not precluding brilliant theoretical work, has had the effect of reducing interest in the general body of research and of hindering communication with overseas scholars, many of whom no longer speak the same technical language. It also considerably weakens the position of geography *vis-à-vis* other sciences, since it appears as little more than a rudimentary discipline that will inevitably be broken down and absorbed by the more systematic subjects, and as one that offers only an outline framework, colourful regional descriptions, and a few case studies as the basis for theoretical speculation, which geographers have proved incapable of handling themselves.

Explanation and models

Techniques of explanation are a critical problem for geographers, who claim to account for as well as to describe selected features of a landscape. Following generally accepted theories in scientific philosophy, many geographers have turned to the concept of causality, but the role of causal explanation is itself the subject of heated debate. There are those who concede its relevance to the natural sciences, but dispute its value in the social sciences where they believe relationships can only be established with varying degrees of probability. However, contemporary research in sociology would seem to indicate both the possibility and the necessity of adopting causal concepts, and geographers would be well advised to consider this issue in greater depth. Until there has been sufficient research, we can only pursue a rather naïve empirical approach.

It is true, however, that much geographical discussion has implicitly or even explicitly turned on this ill-defined concept of causal relationships. It was carried to extremes by Huntington and his disciples (1927), who saw all human activity as being totally dependent upon the physical environment. Other geographers adopted the more flexible position of possibilism, including Vidal de la Blache, who was influenced by the lively interest in probability theory among French mathematicians and philosophers at that time. We must however qualify this relativist viewpoint in geography: the same underlying elements, whether attributes of the physical environment or a pre-existing complex of human factors, do not always give rise to the same consequences, and all consequences are themselves modified through time, that is, they are evolutionary.

A further problem arises from the complexity of geographical facts. Relationships are rarely simple and direct: they more usually comprise a system of interrelationships, with direct and indirect consequences, and recurring or cumulative effects that are difficult if not impossible to measure.

An industry, for example, is established in a particular town because it offers certain essential facilities (the availability of a labour supply and a market . . .) but the prosperity of the industry in turn stimulates urban development. The urban population is increased by immigration, and since migrant labour is usually of the child-rearing age group, the growth rate itself increases. This produces a snowball effect. A subtle interweaving of cause and effect develops from an initial stimulus which is often difficult to define – in the above example, is it the construction of the actual factory complex or a temporarily favourable urban environment which attracted the attention of the firm's managing director? The establishment of the same complex elsewhere would perhaps have had little impact, and without this industrial development the pattern of urban growth might have been very different but ultimately have led to as great a prosperity.

In any discussion of empirical methodology, we should remember that research into explanation in geography has two major dimensions – space and time.

One of the legitimate objectives of geography is to explain the location of a spatial element, despite the fact that its origin is often uncertain. A textile mill, which is now situated in a large town and imports all its raw materials over considerable distances, may owe its existence to a tradition of local craftsmanship, using the wool from sheep kept on nearby pastures (Roubaix and Tourcoing in France), or to the preservation of close ties with a distant metropolis, which is both progressive and technologically advanced and also the source of a skilled migrant labour force (the origin of the textile industry in New England), or to a particularly dynamic social group, often comprising recent immigrants, who at the right moment can supply both the driving force and the financial backing to develop a new resource base (as in Brazil, where prosperity based on coffee led to the expansion of São Paolo).

A single event or a whole chain of interrelated factors therefore account for the present industrial landscape. The present is almost always explicable in terms of the past, either as a process of continuation or as a counter-reaction – the 'discontinuous phenomena' discussed by Brunet (1968). To the extent that a break occurs as the result of previous developments, however, it may still be considered a consequence *a contrario*. For example, modifications of the land use system may be caused by soil exhaustion resulting from long continuous cultivation of the same land (coffee in the state of São Paolo), by a transformation of market conditions with disastrous repercussions on the traditional agricultural system, by an exodus of the younger elements in the rural population, depleting the agricultural labour force, and so on. Consideration of the time dimension in research into the present landscape requires further elaboration, since it highlights the differences between geography and economic history. The economic historian tries to establish the full sequence of events through time: all factors are therefore relevant, including details of the general economic situation and economic trends at a given time, and of a whole range of immediate consequences such as a localised famine or a temporary modification in transportation flows. He reconstructs a local situation in the light of the vicissitudes of time and space. Whenever one is concerned with the influence of past events on developments in a particular time period, whatever the events they belong to the realms of history.

The geographer's objectives are different: he is not primarily concerned with tracing variation through time, and considers past events only to the extent that they help in explaining the present. Economic developments which had only immediate consequences, without trace in the present landscape, are the province of the historian. From a long and often complex process of evolution, the geographer must select only factors which demonstrate that a particular element is not an isolated occurrence, but one link in a long chain that must be patiently re-assembled. This involves two delicate operations. Firstly, in order to choose the salient factors he must clearly understand them and be in a position to judge: logically therefore the geographer should himself undertake research into economic history or, perhaps more simply, be able to interpret the results of those who have done so — this once again illustrates the unique position of geography as a discipline. Secondly, all choice by definition implies a process of elimination, which itself presupposes a sound judgment based on a full understanding of present conditions and problems, and on great skill in the backward reconstruction of linkage mechanisms.

This gives rise to a further problem: what are the limits to this retrospective research? Flatres (1957) demonstrated that certain agricultural features of the Celtic fringe areas of the British Isles (Ireland, Wales, Cornwall, Isle of Man) date from the time of the earliest indigenous settlement. Nougier (1959) suggested that soil degradation on the plateaux and plateaux margins between the Loire and Seine valleys is attributable to a system of cultivation in this area going back more than 4000 years. In countries with a settlement pattern of great antiquity, the search for explanations goes beyond history to examine prehistoric evidence. Stones and fields preserve outline traces of human activity for thousand of years. It is not only rural areas which require the analysis of such remote data: numerous town sites in the Orient and even in Europe were founded several thousand years ago.

The momentary equilibrium or disequilibrium between environmental conditions and human activity must also be reconstructed, since this sets in motion a new cycle of relationships and launches a period of revolution, both of which culminate in the complexity of the present.

What empirical procedures should be adopted? The safest method is to begin with a description of the present, since this avoids the inevitable temptation to digress, and to look for historical influences only where they might fill gaps in the explanatory sequence which emerges from the detailed preliminary analysis. On the other hand, the nature of the data to some extent imposes its own framework on the later stages of analysis — though not on the actual processes of reconstruction — when one of two chronological procedures can be followed. One can either consider the most remote factor and then all other significant events in sequence, thus establishing the processes of evolution up to the present day or, like Flatres, work backwards through time and steadily dismantle all the mechanisms right down to the initial stimulus.

In neither case, however, should one confuse procedures for analysing evolution with research into the causes underlying this evolution. Confusion between causes and interrelationships is more easily avoided where there is a clear temporal sequence of events, such that A → B but B ↛ A are self-evident, but far less obvious when the elements are contemporaneous. For example, the construction of a bridge across the Rhône at Lyon led to the rapid expansion of that city during the thirteenth century and to the widening of its contacts with surrounding areas – but why was the decision to build this bridge taken in the twelfth century? Why did the need for such a bridge arise? Why was its technological feasibility investigated at that particular time, although it proved a difficult operation and its realisation was delayed? Was there a single cause or a series of interrelated factors? If there was a single cause, that is, if the economic expansion of the city was not already under way, what was the cause of that cause? Confusion between causality and interaction can lead to diametrically opposite conclusions when factors operate simultaneously.

What practical conclusions can be drawn from this discussion? As a first step, we must clarify our hypotheses and be more rigorous in establishing correlations. This is relatively easy where the factors involved are in a time sequence, but much more difficult where several causal relationships operate at the same time and are themselves heterogeneous.

This problem arises in all the social sciences but more especially in geography. A geographical feature such as an agricultural landscape derives its own characteristics from the physical environment (itself a complex system of interrelated phenomena), from processes of historical development and from the agricultural systems of successive human communities. Its present appearance expresses the temporary equilibrium between all these influences, that is, a balance between negative forces (the constraints of the environment, the legacy of the past, the lack of dynamism among the present inhabitants) and positive forces (environmental potential, the results of man's dedicated hard work in the past, and the diverse abilities of the population today).

Geographical explanation hinges on the study of the ecosystem and of ecological relationships – 'functioning and interaction between one or several living organisms and their total environment, both physical and biological' (Forsberg, 1962). Within the basic framework of the ecosystem there are two aspects of interest to the geographer: the influence of the past and of evolutionary processes (genetic relationships) and the importance of current interaction (functional relationships). The ecosystem obviously has spatial expression since it is itself a fundamental spatial unit.

The systematic analysis of relationships is considerably simplified and structured, and is therefore more productive, through the use of models.

Models are becoming the panacea for a number of disciplines and this fashion has also caught on in geography. Many foreign geographers are using sophisticated and precisely-defined models while others, particularly in France, adamantly refuse to acknowledge their relevance although sometimes, grudgingly, they are forced to make use of them. A third group uses the word model to cover almost any form of calculation. A review of synonyms (Cole and King, 1968) shows that the term embraces a wide range of concepts and that several definitions are often suggested by the same authors: a frame of reference, a description, a theory, a proposed method of research, a representation, an abstraction, a formalised or semi-formalised theory (*theoruncula*). Harvey (1969) refers to numerous publications in his list of the different types of model: models which represent processes of interaction, which are classificatory, which suggest comparisons, and which are directed towards research into new theoretical formulations or illustrate a pre-existing theory. The eclecticism of their collaborators also forced Chorley and Haggett (1967) to acknowledge these widely divergent views.

To my mind the most acceptable definition was suggested by Braithwaite, who also coined the word *theoruncula*: 'a formalised theory . . . more modest than a theory' (1960). From this conceptual confusion we should salvage the basic idea of a systematic structure, capable of leading to theoretical propositions and assisting their empirical verification with reference to clearly-stated hypotheses. This structure can take many forms: graphs, maps, mathematical formulae, hypotheses.

A model is not some wonderful talisman but a rigorous and effective framework. 'Scientific models are utilised to accumulate and relate the knowledge we have about different aspects of reality. They are used to reveal reality and – more than this – to serve as instruments for explaining the past and present, and for predicting and controlling the future' (Ackoff, 1962). Harvey shares this opinion: 'in the absence of firm geographic theory, a model can provide a "temporary" explanation or an objective (if often inaccurate) prediction . . . But in terms of basic research the primary function of model-building in geography must be directed towards the creation of geographic theory' (1969).

A model is useful at several stages in geographical explanation, enabling us to clarify our thinking and therefore to produce more scientific statements. It is crucial to the overall progress of the discipline for three reasons. Firstly, explanation is dependent on comparison and a model can supply the essential framework for such comparison. Secondly, if the degree of correspondence between the model and the results obtained from numerous case studies is measured, a truly geographical theory can then be formulated or an analogy established with an existing theory in a related discipline, and its application to geography clearly demonstrated. Finally, more simply, a model offers a framework within which to present research findings and therefore facilitates the comparison of results. In cases of discrepancy – and we must not confuse the farmer with his plough: the model is the tool and nothing more – it must be modified so that a sounder scientific basis to our discipline is established.

Three types of model are therefore relevant to geography. A *reference* model is essential to statistical comparison. This is clear from maps of the size of agricultural holdings by municipalities in the *Atlas de l'Etat de São Paolo* (Libault, 1970). Five classes were delimited and shown on a summary diagram by means of a divided circle. The same calculations were then repeated for every municipality: each class with a higher percentage of holdings than the state mean is shown as extending beyond the circumference of the original circle, while each below average class has a shorter radius.

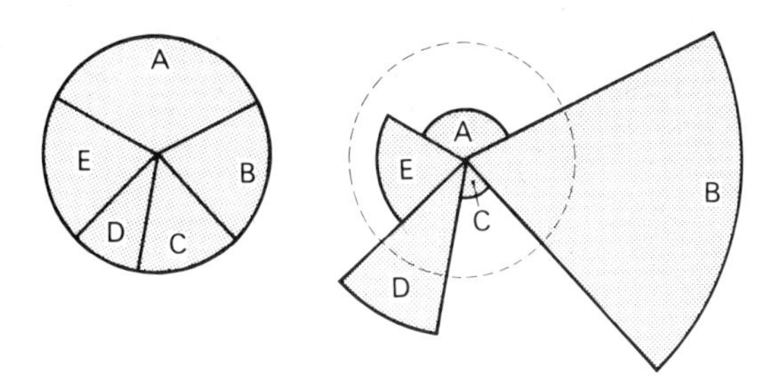

Fig. 2.1 Diagrammatic comparison
Left: *Distribution of land use types in country x*
Ax = 35%, Bx = 20%, Cx = 15%, Dx = 10%, Ex = 20%
Right: *Relationship between distribution patterns in countries x and y*
Ay = 15%, By = 50%, Cy = 5%, Dy = 15%, Ey = 15%
The angles at the centre of the circles are unchanged, but the radii delimiting each sector are proportional to the ratio of the corresponding data. Thus

$$\frac{Ay}{Ax} = \frac{15}{35}, \frac{By}{Bx} = \frac{50}{20}, \text{ etc.}$$

Explanatory models determine the sequence of theoretical steps in research and experimental work. A logical system is developed on the basis of several accurate and highly detailed case studies, and can then be applied to all analogous situations. A study of the demographic evolution of an area requires parallel analyses of natural growth rates and of net migration, tracing the sequence of causes and consequences and their permanent repercussions. A separate analysis could well be carried out for each area, but the criteria selected would probably vary and the statistical calculations and conclusions would not be immediately comparable. The construction of a precisely defined model is of obvious value.

Finally, a summary of both theoretical procedures and results can be presented as a *conclusive* model, which is similar to Braithwaite's *theoruncula*, and which leads to a greater depth of understanding and to new theoretical propositions. The actual presentation can take any of the forms discussed previously, but the geographer should give priority to the development of advanced cartographic techniques (Lehmann, 1968).

Precision – towards a new geography?

In the preceding pages I have several times stressed the need for precision – in observation and analysis, in deduction and interpretation – which leads inevitably to a discussion of the introduction of mathematics into geography, the so-called Quantitative Revolution, of which the most important effect has certainly been 'to force us to think logically and consistently where we had not done so before' (Harvey, 1969). This constitutes a radical modification of our approach to geography. 'In their anxiety to be geographical, geographers have too often condemned themselves to be descriptive, and indeed much of the excellent work now emerging from the application of mathematical techniques to the analysis of distributions is merely a refinement and sophistication of geographical description' (Brookfield, 1964).

This movement is a recent phenomenon and as yet has made little impact on the French school of geography. A book such as *Le fait urbain en France* (Pinchemel, 1963) aroused indignation in a number of scholars, who criticised the aridity of a purely statistical interpretation. Yet Pinchemel did not apply complicated formulae or a bold new methodology, but simply a comparative classification. Some mathematical training, albeit optional, was introduced three years ago for geography students at the University of Paris, and manuals of applied mathematics are now in more general use both in the social sciences (Barbut, 1968; Azoulay and Fried, 1968) and in specialised areas within physical geography. But the application of statistical techniques is still very limited and at an elementary level: at most it comprises the inclusion of some numerical data, but the criteria for their selection and methods of sampling are often extremely rudimentary.

Techniques are far more advanced in overseas countries and in the preface to *Theoretical Geography* (1966) Bunge lists the pioneers of 'the new geography'. All acknowledge that Christaller was the instigator of this revolution and that its success was assured essentially by the Americans and Swedes, as a case of 'independent development at several places at once' (ibid.).

In the United States, the meticulous work by Hartshorne represented the culmination of early American attempts to define geography as a coherent and autonomous discipline (Mikesell, 1969) and to establish that 'geography . . . was an idiographic rather than a nomothetic science' (Harvey, 1969). Since that time almost all geography in America has moved towards the adoption of a normative methodology, no longer founded on apparent and approximate relationships but on sound statistical and mathematical principles. The most rapid advances have occurred since the early 1950s, as geography benefited from the experience of the physical and biological sciences in applying techniques developed by physicists and mathematicians. The real quantitative revolution began around 1950, reached its zenith in the period from 1957 to 1960, and in America is now considered to be over. 'An intellectual revolution is over when the revolutionary ideas themselves become a part of the conventional wisdom' (Burton, 1968). After many years of fierce opposition, this is now the case in geography since quantification is widely accepted, particularly by younger scholars. Perhaps the clearest proof is that courses in quantitative geography have been introduced into the examination syllabus of almost all North American universities. It is also significant that in 1959 Hartshorne, the staunch defender of traditional geography, modified some of the views he had expressed twenty years earlier in the light of these new developments.

Quantification in geography was indirectly stimulated by the work of certain mathematicians, physicists and economists, who had criticised the subject for its lack of precision – as in the article by Stewart on

'Empirical mathematical rules concerning the distribution and equilibrium of population' (1947). Widespread and bitter discussions ensued between disciples and opponents of the new methodology, in which not even the great names of the past were spared (there was a strong attack on the descriptive explanations of W. M. Davis) and irrespective of national frontiers (Wooldridge was quick into the arena to defend the Davisian point of view). The opponents of quantification put forward the same arguments as are currently heard in France, where the revolution is at last under way: that it is both a dangerous and wasteful process since it retards the training of geographers, who must already acquire at least an elementary understanding of several neighbouring sciences and of cartographic techniques, and who can do without additional grounding in mathematics and statistics (a line taken by Stamp in particular, 1957); the impossibility of measuring geographical variables because they are too complex or too numerous; the incorrect and indiscriminate application of the techniques; the inadequate qualifications of the would-be quantifiers, resulting in exaggerated claims for a methodology that is otherwise of considerable interest.

This stage of apprehension or of outright condemnation is a thing of the past, however, and new developments can now take place in a more objective and less inflammatory atmosphere. First results are positive and demonstrate both the parallelism between research in geography and in long-established sciences such as physics, and the possibility that social scientists can continue to observe chance factors at the microscopic level, while seeking to formulate general laws at the macroscopic scale. We must therefore look beyond unique facts for a classificatory theory that will allow us to reach valid conclusions through deduction and comparison, and so accelerate progress towards a scientific geography.

In England, as we have seen, the initial reactions of the elder statesmen of geography were predictable and conservative. But a new generation of scholars is rapidly establishing itself through publications in two fields: technical considerations (Haggett, 1965; Chorley and Haggett, 1967; Cole and King, 1968) and epistemological questions (Harvey, 1969). As in the United States, the one is seen as the logical extension of the other, and both clearly illustrate the importance of quantification and of new research methods. From 1953 onwards the school of geography at Lund published a series of exploratory essays, including the one by Bunge which caused a considerable stir. It was a purely theoretical work as 'there are many books on geographic facts and none on theory' (Bunge, 1966). We must redress this balance in order to convince all geographers, whether mathematically inclined or not, of the crucial need to demonstrate the scientific basis to geography.

Developments in the USSR have partly reflected the course of political events. A chair of economic geography was established at the University of Moscow in 1929, with the appointment of Professor Baranski. Although this branch of geography was brilliantly taught in several universities and was the first to show signs of progress in quantitative research, a survey of systematic geography in the Soviet Union in 1960 still made no mention of statistical methods. The first articles to consider them seriously appeared in conference proceedings published in

Moscow in 1964. They were summarised for general circulation in an article by Vasilevskii entitled 'Mathematical models in economic geography' (*Sovetskaya Entsiklopediia*, 1966, vol. 5). Anuchin's book (1960) had already marked the liberalisation of Soviet economic thinking after the death of Stalin. In the article by Vasilevskii, economic geography is defined in several ways but the need for practical applications to economic planning based on Marxist principles is always paramount: 'the economic structure of a society determines the superstructure of society, that is, the ideas, institutions and organisations of society'. His reference to mathematical methods is to mathematics in the widest sense: 'a distinctive symbolic system which avoids polysemantic ambiguity and permits abbreviated and interrelated statements'. It is more important to have a general understanding of the system than to obtain concrete numerical values for the behaviour of variables in the system: 'the principal aim in building mathematical models . . . is not so much to obtain numerical results, which are often of doubtful value because of our inadequate knowledge about basic constants and functions that enter into the problem, but to determine the actual structure of the solution to the problem' (Jensen and Karaska, 1969). Soviet authors also caution young workers against numerical empiricism, which is alien to theoretical formulation and offers no possibility of discovering new laws. They stress the need to apply mathematical techniques that have been tried and tested in other disciplines, and for research into laws and theories rather than into verification procedures for the basic statistical data.

This lively interest in quantification in the major schools of geography is reflected in the current flood of books and articles on methodological issues, in the adoption of mathematical tools for geographical analysis, and in the use of conceptual models (economic regions, transport networks, systems of urban hierarchies, etc.), which are now indispensable to progress in comparative research. Without the use of such models, we cannot reasonably claim that the geographer has any contribution to make to the delimitation of planning regions, an essential stage of any genuine national policy for resource management.

It should not be assumed from this brief review that the way ahead is now quite clear, nor of course that the new trends should replace all inherited methodology. Many geographers use the terms 'quantitative geography', 'mathematical geography' and 'new geography', but again there is need for caution and precision since each author does not necessarily intend the same interpretation. Worse still, there are a number of geographers who claim that analysis based on figures and formulae constitutes the only valid research, or even that it is the only salvation for geography – as though econometricians were claiming to be the only true economists! It is dangerous to prolong this confusion for, on the pretext of being innovative and scientific, we shall deal a death blow to geography.

Three definitions recur in the many publications from both the Soviet Union and the USA, each corresponding to an aspect of geography that is itself undergoing a major transformation.

Quantitative geography is simply geography which makes use of precise measurements, whether classificatory or of a single variable. At its simplest, it has long formed part of the accepted methodology of most geographers, and consists primarily of replacing a more or less vague adjective with a numerical value. This can ultimately lead to classification and comparison and to the establishment of interrelationships. Harvey distinguishes several types of measurement: nominal scaling (1, 2, 3, . . .); ordinal scaling – ranking objects or events in order of magnitude, with or without a natural origin; the measurement of distances between two objects on a scale, either by interval scaling with an arbitrary origin (for example, location by coordinates on two axes where the point 0 is a precisely-defined common origin) or by ratio scaling with a natural origin. Quantification obviously demands more than simply the inclusion of raw data: it requires the use of mathematical tools such as logarithms, trigonometry, derivatives and integrals.

Two points arise from this. Firstly, all measurement is far more complex than might first appear. To determine distances, for example, might seem child's play but great care is needed in defining the point of origin – for a town, should this be an officially-recognised focal point such as Notre Dame for Paris, a geometrically-defined centre, or a point on the outer periphery? Which distance is to be measured – as the crow flies, by road or by rail (all of which follow different routes), ground distance, or a time or cost measurement? If we opt for cost–distance, should we simply take into account the total sum of money spent on transport, or also the time lost or gained by following an alternative route that is less arduous but longer? This is not a frivolous question but reflects a fundamental problem in geographical research, arising from its study of the complexity of the total environment. How for example does one measure the rentability of a rural estate, levels of living, or stages of economic development? The simple determination of valid criteria, let alone their subsequent manipulation, is a delicate operation and one that is obviously more challenging than brilliant subjective interpretation, satisfying in its literary allusions but lacking a scientific basis.

The second observation concerns the mass of statistical material acquired during the quest for the quantitative.

> These numerical data should be analysed by sound statistical methods so that maximum value is obtained from them. Too often a considerable body of valuable quantitative data is presented either in a raw state or after a minimal amount of processing . . . more fundamental and possibly more valid conclusions could be reached, or varied aspects of a problem investigated, by means of a more comprehensive and subtle use of existing statistical methods (Gregory, 1968).

The second aspect of the new geography is *statistical geography*. This goes beyond initial measurement to the stage of using a standard set of procedures to analyse interaction, to establish comparative relationships and to verify hypotheses. The selection of data (sampling techniques), calculation of descriptive statistics (mean, deviation around the mean), comparison of sample values (correlation, analysis of variance) and estimation of margins of error and of probabilities are the major steps by which geographers establish a firm foundation for their hunches and intuitions. They must also learn to be sceptical of incomplete or unreliable data, of superficial similarities and of biased sampling, all of which lead to errors in interpretation.

Finally, the third aspect is what might properly be called *mathematical geography*. As with the other two, it cannot oust all traditional geography and like them it is an additional tool. Its major contribution lies in the clear formulation of basic principles of logic, expressed through a distinctive language and system of reasoning. 'Mathematics is the art of describing different objects by the same name . . . When the language is well-chosen it is amazing how the designation of a familiar object is suddenly applicable to a whole new range of objects: nothing need be changed, not even the words, for their names have become the same' (Poincaré, 1908). When the geographer embarks on explanation and comparison, his theoretical framework will be more rigorous if it is expressed in the language of mathematics: it will be cumulative (referring to comparative norms obtained from previous analyses) and communicable (intelligible and applicable by other research workers, and by implication therefore more scientific and productive). Part of this expression will be as formulae symbolising the relationships between structural and functional variables. An appropriate language is indispensable, the mastery of which is often either neglected or erroneously seen as an end in itself. The following model illustrates two possible approaches to problem solving and the role of mathematics in geographical analysis (Harvey, 1969):

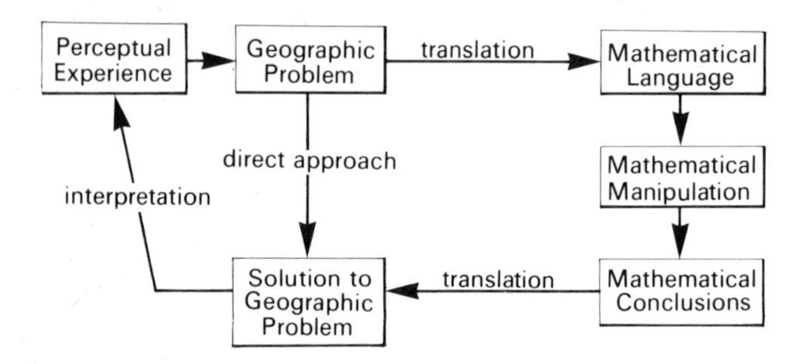

Fig. 2.2 Problem solving and the role of mathematics

'Mathematics does reveal its full usefulness to geography since the mathematical framework of theory is used to carry geographic thought beyond intuitively obvious results obtainable with descriptive statistics' (Bunge, 1966). Of the three developments in modern geography, that of mathematical geography is probably the most significant. Firstly, it must by definition underpin the other two if we are to order and classify data, and to extract from them the maximum value in reaching sound conclusions through logical deduction. Secondly, mathematical geography is internally consistent: in cases where numerical measurement is impossible, respect for logic still allows us to develop a satisfactory theoretical framework.

Perhaps it is not too great a simplification or overstatement to suggest that quantitative geography is descriptive, statistical geography is correlative, and mathematical geography is conclusive! The three are not of course mutually exclusive: their possible applications are interlinked and together they constitute the essential scientific basis for geography. Certain branches of the subject lend themselves more successfully than others to this new treatment, and not all rely on the same techniques. Relevant to spatial analysis, for example, are geometry (Bunge, 1966), network analysis (Ullman, 1957), factor analysis (Berry and Marble, 1968; Berry, 1968) and topology.

The application of all these techniques has greatly improved through use of the computer. Its capacity to store information, the accuracy of its memory and speed of calculation allow long complex programmes to be run through in a few minutes whereas a research worker, even with a team of assistants, could not produce the same results within a reasonable period of time. A further development is its potential application to cartography – not as an alternative but as a supplementary tool. At the present stage of technological progress, careful design of the general-purpose map should still be undertaken by the geographer–cartographer while the analytical map, reproducing in a few seconds the comparative distributions of ten or a hundred spatial factors, is the work of the computer. The computer also verifies hypothetical curves and predicts the outcome of evolutionary processes, or of an evolution modified by the introduction of new factors.

An excellent practical example of the use of the computer is our recent work on the *Atlas de Paris et de la Région Parisienne*. Paris consists of 5000 districts, and at the time when the maps were drawn no records of demographic data – population, employment, etc. – existed at this scale. We therefore had to start from scratch and the maps of household amenities or of the average number of persons per room, which took three months of survey, calculation and manual cartography to produce, can now be reproduced in fifteen minutes once the computer has been programmed to recall the variables collected and stored at the district level – and with further progress in automated cartography this will take only a few seconds. The time required to complete the work on the Atlas today could be reduced by 80 per cent, and if the data were collected for uniform spatial areas as in the USA and Sweden, by 90 per cent. Data on sixty variables relating to 3 million farms are now available in Sweden: an appropriate programme will reveal in a few minutes the location and yields of farms of a given size, the type of farm buildings and cultivation pattern . . . or even the effect of legislative measures on certain farms, and the income of farmers from a specific age group and region of origin. In the United States, where data for urban areas are recorded by blocks, an atlas of twenty large American cities was produced showing strictly comparable distributions of population, dwellings and income level . . .

Discussions of the merits and limitations of the use of a computer have appeared in numerous publications, and the various arguments were summarised by Kao (1968). Additional references include articles by Salitchev (1970) on basic socio-economic maps which can be produced in all their complexity only with the aid of modern techniques, and by Gould (1970) and Spence and Taylor (1970) on the role of quantitative methods in spatial analysis.

Thus the wish of many scholars, expressed by Mill at the turn of the century and cited by Freeman (1961), can now be fulfilled. 'Geography as a science is the exact and organised knowledge of the distribution of phenomena on the surface of the Earth . . . the attitude of geographers has hitherto been directed mainly towards the collection of facts: we now require to discuss and arrange them.' However, he added somewhat dejectedly after the publication of his model study of southwest Sussex that 'for one man without funds to give a true sample of a work that required a trained staff and free access to all necessary data was an impossibility'.

Research in geography is increasing in scope and momentum, but can only benefit from this superb new technology if geographers apply themselves seriously to familiarisation with quantitative techniques. A computer can only manipulate the data fed into it: there will be no output if it receives no instructions or only garbled messages. But when carefully programmed, it can throw new light on structures and relationships that were only partially apparent or suspected on first inspection of the data, or could be identified only through laborious longhand calculations. Demographic research for example, for which a large quantity of census material is available, is already making rapid progress. The current problem for geographers is to collect or extract sufficient comparable data with which to feed the computer and so verify their hypotheses. If sufficient data were available at a small scale (the commune in France for example), regional geography might regain some of its former glory. A detailed example of this possibility is seen in the article on 'Regionalisation of Pennsylvania counties' (Stevens and Brackett, 1968), where despite the relatively large area of individual counties the value of the technique is clear. Gould (1970) also demonstrates the speed and precision which the computer brings to spatial analysis.

We might end on a light-hearted note. According to le Littré the computer 'restores order to chaos': if the geographer 'seeks for the hidden order in chaos', is not a liaison inevitable?

Geographical method and scientific method

Both in research techniques and in the presentation of results, geographical methods have undergone radical modifications, acceptance of which has already occurred in numerous countries and is imminent even in France. The trend is towards more precise measurement and definition of a more rigorous and quantifiable theoretical framework. We must not harbour illusions, however: numbers are not science and quantification will not in itself, like the wave of a magic wand, earn geography a place in the ranks of the recognised sciences. Accurate measurement and the use of more powerful analytical tools increase the chances of progress, but they also increase the risk of damage through mishandling.

'The quantitative revolution implied a philosophical revolution' (Harvey, 1969). Geographers must know not only what steps to take, but also where they are going and why they have chosen a particular route. To travel is to arrive – confirming the reality of geography at the same time –

but we must agree on our ultimate destination. At several points along the way there will be a turning or a choice of paths, but we must continue straight ahead or risk getting lost. But how can we go forward if we have no idea where or why or by what method we are travelling? We must continually check our sense of direction.

The question of whether or not geography is a science is often disputed because of its apparently unpredictable, shifting and ill-defined field of study. Is it the objectives or the methodology of a subject, however, which make it a science? One might also consider whether or not this idea that the field of geography is complex is an exaggeration in the mind of the individual research worker. 'There is no doubt that everything can be an object of scientific investigation. The possibility of a science existing is determined not by the what but by the how, not by the subject but by the method' (Korach, 1966).

It would be regrettable if geographers, particularly in France, were to be distinguished by an arrogant refusal to participate in the great scientific revolution of our time. Perhaps we should try to reconcile Piaget's statement that 'a science which has no structure, no objectives, does not exist' with Sauer's claim that 'a healthy science launches into discovery, verification, comparison, generalisation: its objectives will be determined by its capacity for organisation' (1963).

3

Geographical space

The geographer's space

A complex space

A geographer studies the relationships between man and environment, between two visible components of terrestrial reality. But the concept of relationships is itself complex and therefore difficult to translate into simple spatial expressions. It comprises both visible features and abstract processes. The peasant ploughing his fields, the development company clearing a wooded site for new housing, the planning organisation directing the excavation of a canal, a dry climate necessitating large-scale irrigation works . . . all are situations that are immediately visible and concrete: they explain or translate direct relationships of adaptation or transformation between man and his environment, and obviously fall within the scope of geographical study. On the other hand, the large motor company with its headquarters in Detroit deciding to establish a subsidiary firm in Germany, feminine fashion infatuated with Indian designs causing a massive importation of scarves and cloth from Bombay into western Europe, the cheapness of Italian electrical goods enabling them to capture the French domestic market, the deficit in foodstuffs in a particular Asian country giving rise to waves of cereal imports from various parts of the world . . . also fall within the range of geographical considerations. Decision-making, forces of attraction and competition are all terms which have geographical connotations although at first they may seem to belong to sociology, economics or some other discipline. The decision of the board of the American firm will be translated into the choice of a suitable location, the construction of a factory, the recruitment of a labour force and research into conditions of supply and demand. Economists, sociologists and geographers are all involved with the whole series of operations, each pursuing his own approach and examining every stage from his own viewpoint – for example, costs and the market for the economist, motivations of the labour force for the sociologist. The geographer should normally take as his starting point a feature inscribed on the Earth's surface, such as the factory complex. He is concerned with one particular stage expressed in a material space and with the reconstruction of the preceding and subsequent links in the chain: if possible he should explain or at least identify the sequence of events. Similarly, the appeal from the areas of India suffering from food shortages (in this case the significant local relationships are between the size and technological capacity of the population and environmental possibilities) is translated into the export of foodstuffs from countries where a surplus is itself the expression of geographical relationships: commercial flows imply cargoes, ports . . .

The first concept which can therefore be distinguished is the operation of a dual system: firstly, the existence of highly diverse relationships, the elements in which are themselves variable and widely distributed over the Earth's surface; secondly, a concrete localised expression characterising these relationships at one moment in time and within a clearly-defined framework. Biological, physical, sociological and economic factors all influence the nature of the relationships, a hazardous complexity to analyse, particularly at a later stage when the geographer claims that his is the science of synthesis. Surely he has overstepped the limits of his field of study? 'The nature of perceived space and the limitations to man's thought processes indicate the difficulties to be overcome in defining his role. An object used for a specific purpose, and precisely defined both semantically and within the context of a research project, may belong to numerous spaces simultaneously' (Perroux, 1954).

The geographer must find his niche among the many dimensions, for it is 'geographical space that underpins systems of relationships' (Dollfus, 1970). His field of study lies at the point of contact between these complex and diverse relationships and the Earth's surface. It is both the localisation of economic space, the material expression of psycho-sociological space, and the contemporary expression of historical space.

This statement is crucial to any assertion of the uniqueness and relevance of geography, which should not merely be an unsatisfactory 'inventory of economic units within some container space' (Perroux, 1954). Geographical space is not simply a more or less deterministic physical space, nor an analysis of limited areas of the Earth's surface. Even in a material sense, location in geographical space is more apparent than real since it is an isolated coincidence in one fragment of the total environment. One observes a single point of convergence between facts that are widely distributed in both space and time. The morphology of a valley is a function of the distance from and the nature of the source of running water, of its evolution through successive geological periods, of its catchment area; climate is related to the movement of air masses. How many factors must therefore influence interdependent economic units?

A concrete space

On the other hand, no branch of geography should comprise merely theoretical or statistical speculation on exchange flows or production figures, as has happened for too long with certain types of economic geography. This would betray the objectives of the subject and produce only second-rate economics or mediocre statistics. Its aims are very different and its field of study, while extensive, has one fundamental characteristic – it is firmly rooted in concrete space. However it is defined, the geographer's space does exist. It is not a convenient abstraction like economic space, nor bounded reality like geometric space. It is both real and invisible, fixed and moving, and is more than simply a question of location. It presents constraints and opportunities which cannot be ignored: facts are simply located at a single point or on one particular plane. All relevant data must be precisely understood and clearly illustrated: the use of photographs and maps reflects this major preoccupation of our discipline. Maps in particular allow the geographer to project in a meaningful way and on a flat surface a high proportion of the data derived from total space. Commercial flows, links between distant commercial headquarters and nearby factories, the position of air masses determining a local climate – all kinds of relationships and evolutionary processes can be represented on a simple sheet of paper by means of appropriate symbols.

A geographer must consistently focus his attention on material space, on the projection of an abstract concept into a tangible reality. Visual observation is the initial process. Take the simple case of a relief form, an unusual land use pattern, or a radial network of roads around a town: the geographer perceives space at two levels – firstly, general and direct observation common to everyone considering the same feature and secondly, the perception of a specialist who develops his observations into a problem for investigation. But why does this second stage exist and why in this form? For the geographer, space is both static and active. He responds to the movements of his eyes as they survey and probe it, and the mental image which forms is not complete, finite or passive: it automatically provokes a response mechanism (Downs, 1970). The features inscribed on the Earth's surface and the fact of observation are the incentives for continuing research into correlations, in an attempt to reach a comprehensive explanation.

Direct observation, examination of aerial or of ordinary photographs, study of detailed cartographic records, or even reading accounts by a third person (for example, the narrative of a voyage or the description of research carried out by specialists in other disciplines) may all constitute the starting point for research and directly, or through a lively imagination, provide the visual stimuli. There is no substitute for this preliminary stage which is one of the unique features of geography – the observation of a concrete fact.

Many geographers believe that the real basis of their discipline is the landscape, which 'unites the gifts of nature and the contribution of man' (Labasse, 1966). But what is this landscape? Geographers of the French school 'have described the landscape but avoided all controversial issues' (Sorre, 1957) – hardly an epistemological approach but one which for some time proved both dynamic and productive. It now involves so much ambiquity and obtuseness, however, that it must be re-examined.

Specialists in Germany and America in particular have questioned the meaning of the word 'landscape'. Hartshorne (1939) devotes many closely-argued pages to the issue in his long dissertation. Several American authors have used the term without specifying the precise meaning they attach to it, but often in the sense of a region – also a common interpretation of the German word 'landschaft'. This is the first ambiguity: should landscape be used to refer to a limited sector of terrestrial space or to an area without precise boundaries? Sauer (1927) claimed that the landscape includes both natural features and those superimposed by human activity: the Berkeley school has also adopted the idea of a cultural landscape. It is not entirely satisfactory, however, since it almost always implies visible features, whereas Brookfield (1964) has demonstrated that such a limited interpretation inhibits explanation in depth and also paralyses any interdisciplinary research. While one may admit that physical features often reflect phenomena that are not directly observable such as climate, the effect of which is indicated in the vegetation cover, in the type of erosion and even in certain aspects of human life and activity, this is not true in the same way for all human features, many of which remain without expression in the landscape.

Amid such widespread confusion the comment by Preston James is apposite: 'each geographer who uses the word landscape gives it, albeit subconsciously, a special meaning'. Even in French, the simple term 'paysage' is somewhat ambiguous: as with 'landscape' in English and 'paesagio' in Italian, it suggests essentially an external and therefore a descriptive visual image, equivalent to the 'physiognomy' of Vidal de la Blache, who considered it of such importance that he wrote: 'it is the diversity of landscapes which arouses the sleeping geographer in each one of us'. 'Landschaft' however is far richer in meaning, which may explain the predilection of German authors for a more comprehensive usage (Uhlig, 1970). Schmithüsen considers six accepted definitions of the word in a recent article and concludes that 'the whole of geography is contained within the concept of a landscape' (1963).

The expression may perhaps be modified satisfactorily to 'geographical landscape', for which an excellent definition was suggested at the IGU Congress in Amsterdam in 1938:

> A geographical landscape is not simply a physiological or an aesthetic entity: analyses have shown that it embraces the whole range of genetic and associated functional relationships between units on the Earth's surface, enabling characteristic groups and sub-groups to be identified.

This conclusion is re-iterated by modern authors such as Sestini (1963): 'it is appropriate to use the phrase "geographical landscape" when discussing not simply external appearances but the actual relationships between objects and phenomena'. Ideas similar to the above had already been expressed by Juillard (1962):

> The landscape to the geographer is more than the external reality (which is the landscape in general). Beneath the visible surface he must identify the momentary balance in the underlying relationships between natural features, demographic structures, technologies for transforming the environment, the inheritance of past relationships and new ones in the process of development, types of economy, and social structures and relationships . . .

Juillard thus spells out what was implied by the general definition at Amsterdam.

Is a more precise and clear definition necessary? Juillard's statement is a sound one, but in its comprehensiveness it includes in my opinion two separate and successive empirical procedures. As I have already indicated, the geographer observes the landscape with his eyes and with the perception of a specialist. The study of a landscape involves many stages – not simply description of the visual impact of a more or less extensive area and its cartographic representation, but also the analysis of factors relating to spatial structures that are not immediately visible, and which can be unearthed only in specialised documents such as tables of statistical data. If we examine a rural area, we shall discover directly or with the aid of maps the location, spacing and size of the villages, the shape and size of individual plots of land, the type of land use, the presence or absence of enclosures and numerous other details. Certain indisputably geographical facts will still be missing, however, including both spatial information such as the area of farm holdings and other quantitative data. The same would be true of an urban area. Direct observation of a single landscape would lead the geographer into making ill-founded assumptions, and into proceeding more through guesswork than with a firm grasp of the elements and problems of the space under consideration. A first conclusion is obvious: although the geographer takes into account the components of the visible landscape, he must eventually complete his investigation by studying invisible attributes of the original data – the real problem area of his research. Secondly, it is only after taking this first basic step, this sensitive appreciation of the landscape, that the second stage is possible – 'research into the momentary equilibrium between underlying relationships', using a wide range of analytical techniques (Juillard, 1962).

All the initial elements, either individually or as a group, are then re-examined in relation to the constraints of the physical environment, to the vicissitudes of the past and to human factors. The whole range of temporal and spatial processes are considered and the observed elements are only an underlying thread. One must not lose sight, however, of the need to establish a logical sequence of relationships.

A coherent space

The above heading implies that geographical space should be distinguished by a general coherence. The analysis of observable phenomena is based on correlations that are to a large extent predictable, although the processes may vary in detail: relief is a function of structure in its widest sense and of climate; the plan of a town must be considered in relation to its site, origin, and fortunes of development. Simple 'exploratory models' already exist for use in such analyses. This is not to say that all predetermined stages will occur in any given situation, nor that there is an automatic response, but that there are certain technical guidelines based on the results of previous research from which new leads may occasionally develop. Around 1950 for example, an understanding of periglacial processes and related morphological features led to rapid progress in the search for an explanation of relief forms in regions of the world that are now cold or temperate. This discovery is a clear illustration of the methodology under consideration: the observation of morphological features, searching questions, analysis of the subsoil, laboratory experiments, collation of data collected by numerous research workers at a variety of sites, identification of the mechanics and basic processes of evolution, research into correlations with visible indicators, interpretation of all aspects of the problem and of the logical sequence of correlations. The previously accepted explanations for many morphological anomalies have had to be completely re-interpreted in the light of these procedures.

In the majority of cases, however, a given set of initial conditions does not lead to a single outcome. The development of an irrigation system, for example, is generally accompanied both by an improvement in agriculture and by a rise in population; an exodus from rural areas can lead either to an improvement in farming techniques and in yields or to the abandonment of land; the establishment of a factory or the growth of a town may give rise to population movements, to the development of a transport network, to modifications of the natural environment. The research worker who is experimenting with a new line of approach that may suggest several possible outcomes rightly takes every initial observation into account: progress in his inquiries and time for reflection will then enable him to eliminate certain solutions or to specify them more precisely. There is a logic in geographical research. I have just used the word 'experimenting': this is an extremely important modern development in the subject which has placed it on the margins of – if not already within – the field of sciences. Acceptance as a science is anyway inevitable, just as soon as quantitative methods are more generally applied in all branches of geography, in all countries, and have themselves been still more systematically developed. Geographers in this half of the twentieth century are not working in virgin territory: the workshop for composite sciences such as geography, concerned with both physical and human issues, does involve a certain amount of systematic experimental work but relies primarily on field investigations and the preparation of monographs. Thousands of such monographs on urban and rural areas and on commercial activities, hundreds of studies on the weathering of rocks and on slope development are indispensable raw material in establishing a typology, in formulating generalisations and in drawing conclusions, although there are a few notable exceptions.

Certain geographers resent the idea of quantification and of the comparability of geographical data. It is easy to reassure them: geography will not be transformed into a series of mathematical formulae, nor into a more or less random set of statistics. The geographer is and must remain basically a man of letters, by training as well as in his methods of description and synthesis. The richness of the written word and the subtlety of a personal analysis, almost of an individual intuition, must be his guiding spirit. Neither the most elaborate model nor the most perfect statistic can replace them. However, just as the human skeleton provides a basic structure for the outward beauty of the body, so penetrating geographical analysis must be underpinned by the bare bones of statistics: there is nothing to lose and much to be gained.

Once this position has been clearly stated, quantification of our research should be supported enthusiastically, but quantification only for purposes of comparison. All description, however precise, involves an element of personal interpretation which statistical analysis sets in a comparative framework. Weber stated in relation to the historical sciences that the identification of regular patterns and the establishment of general laws are no more than intermediate stages towards the ultimate goal, that is, the understanding of one particular historical event significant in its uniqueness. There is no reason why the uniqueness of a geographical feature should not be tested by the same procedures.

A variable and changing space

Since the geographer's space comprises such variable data, it is not only concrete and coherent but also clearly distinguishable by its relativity: it is essentially diverse and changing.

This space is variable through time. To take some simple examples: a landscape is not the same in summer and winter, at least in temperate latitudes; a single area may be disrupted by the discovery of mineral resources, by the construction of a new road, or quite simply by a change in traditional farming techniques (the introduction of mechanisation or of new species, or the massive use of fertilisers); a town rarely stagnates and its relationships with the surrounding lowlands are modified for numerous reasons. Furthermore, it is often difficult to delimit the extent and precise date of the modifications.

With advances in technology the same cause may produce different effects. Only during the last decades of the nineteenth century, through the discoveries of the two English engineers Thomas and Gilchrist, was it possible to exploit the iron ore of Lorraine (the famous phosphorous 'minette') and to transform the northern part of the region by establishing a large-scale smelting industry. The construction of bulk ore carriers, however, transporting high grade ore at low cost from Sweden, Mauritania or Brazil, and favouring the smelting of iron at waterside sites – of which the plants at Dunkirk and Ijmuiden are among the most modern and most productive – may cause difficulties in maintaining an inland industry based on the output from local mines. The abandonment of exhausted coalmining regions in the United Kingdom, the Belgian Ardennes, and the northwest Appalachians in the United States, together with a general decline in production from the highly-industrialised nations of the Old World, has created vast areas of dereliction or given rise to local employment crises. On the other hand, the discovery of coal basins in rapidly-industrialising countries – the USSR, India, China – is once again drawing workers from the human ant-hill.

The geographer's space is both unstable over time and also mobile – one of its most important characteristics. Man is not stationary: migrations of all types, whether permanent or temporary, regular or casual, of groups or of individuals, seasonal or daily, involve the greater part of humanity. Migrations themselves express spatial variation, but why do men move from their homes? Because they are prompted by the rhythm of the seasons, for work or for leisure; because there exist both rich and attractive areas, or simply those where employment is available, and neutral or hostile regions with inadequate resources to support all who were born or live there. Population movements are the corollary of the lack of spatial homogeneity. They are matched by numerous other flows – transport, trade, capital, ideas. Everything is in circulation, not only human beings. Employees may live in ten or even a thousand different places but work at the same point; urban dwellers may abandon their town during part of the summer for the sea, the mountains or the countryside. When individuals move their dwellings remain; there is both a fixed landscape and mobile elements. This is a dualism within geography crucial to the subdivision of space.

The division of space, the province of the regional geographer, is the logical outcome of an initial analysis of visible features followed by research into the interdependence of cause and effect. The geographer's space is not only modified through time, but also varies from one direction to another. An observer on the summit of the Aigoual sees at his feet the branching channels of the Rhône delta as they reach the shimmering Mediterranean; in front of him and far to the east towers the wall of the Alps, and if the sky is clear the sparkling of a snow-white brilliance; to his right the horizon is broken by the sombre crests of the Pyrenees; if he turns round the wooded ridges of the Massif Central slope gently away beneath him; he may have the good fortune to see clouds rolling in from the west and rain teeming down on the Atlantic slopes; in the opposite direction, in hollows enclosed by the steep edge of the Cévennes, blazes the Mediterranean sun. Having climbed slowly upwards through the conifers of the northern margins, he can then descend rapidly among the chestnut trees and the chirping of the cicadas.

What images could speak more eloquently than these? They illustrate a diversity of spaces within a clearly-defined framework, but perhaps they are too beautiful and too simple. Once he has reached the plains of the lower Rhône the geographer must no longer be carried away by general impressions: he must identify detailed variations on the ground. But should he observe fixed elements (land use types, the density and spacing of human settlements) or rather be alert to flows and movements, to the life which unfolds over these fixed points? Should he base his division of space on static or on functional elements? One of the major causes of the current debate on the concept of a region undoubtedly relates to these different viewpoints.

Before examining in greater depth this dual approach to spatial structures, it may be useful to summarise the discussion so far. Geographical space is a coherent and universal concept, based on the analysis of an element that is directly or indirectly observable at the Earth's surface at any given moment. It is not a simple space however: it is not merely a part of the Earth's surface, even in a three-dimensional form including volume, but a complex of concrete facts and related causes, implications and consequences that are not immediately observable, such as financial flows or a change in attitudes. To clarify this static–functional dualism, we may say that geographical space is an aggregate of elemental units with complex structures, linked through the interplay of forces among which human activity is often decisive. But the 'essential postulate of the general field theory is that the fundamental spatial patterns that summarise the characteristics of areas and the types of spatial behaviour that are the essence of the interactions taking place among the areas are interdependent and basically isomorphic' (Berry, 1966).

Geographical space and economic space

For several reasons, it is useful at this point to outline the attitude of economists to spatial problems in general and in particular to the space claimed by geographers.

Positions must be defined 'prior to the analysis of economic spaces considered as quite distinct from geographical space' (Perroux, 1954). The converse is also true! For Perroux – and he is not the only one – geographical space is very often explicitly restricted to its physical aspects: 'geographical space conceived as a collection of distances and physical barriers' (ibid.). This is a definition to which of course no geographer can subscribe. Far from being confined to a study of environmental conditions, or even to an inventory of human societies in conflict with their environment, he claims to examine the whole range of interaction between the environment and human activity. To some extent therefore his considerations coincide with those of the economist.

Secondly, economic concepts have exerted a profound influence on the development of geography, paradoxically in view of the scant attention paid by most economists to spatial considerations. In the USSR, for example, current thinking in economic geography was stimulated by translation of the works of Isard, while in France many geographers integrate their work on regionalisation into a framework developed by economists. There is also a wealth of joint research and publication in the United States. Conversely the theories of Christaller, a geographer–economist, have undoubtedly been the object of more productive research and methodological investigation by economists than by geographers, who have been content to apply or discuss his work in a highly empirical way. Elaboration of this example would clearly demonstrate the divergence of approach between specialists in the two disciplines.

Finally, many economists berate geographers for the weakness of their conceptual thinking. They criticise their apparent inability to formulate a spatial dogma, particularly in relation to regions and regionalisation, and therefore ultimately to develop a framework of value in economic analysis.

Even if we are to concentrate on current ideas of economic space a brief historical review is not inappropriate, since it will demonstrate both the recency of these ideas and their closeness to certain views expressed by geographers. This relationship has come about fortuitously, through empirical observations, and through the approach of the individual personalities in question.

Space has been grossly neglected in classical economic theory, which has generally been concerned with 'universal quantities in an economy reduced to a point': modern dimensional theory on the other hand enables us to assign spatial coordinates to these quantities, 'that is, to assign them simultaneously a location and a dimension' (Ponsard, 1955). It is somewhat surprising that 'a problem area in economic research related to spatial phenomena . . . was . . . developed only in the nineteenth and particularly in the twentieth century' (Ponsard, 1958), and that it is still the concern of so few economists. The analysis of spatial phenomena has been far less popular than that of dynamic factors: time has taken precedence over space. The concepts are often therefore vague and imprecise since they are only a secondary consideration, almost an incidental by-product of economic theory. There is no established doctrine, few coherent hypotheses and little overall progress – rather a number of key publications in a series of more or less detailed minor contributions.

Von Thünen is generally considered the first to have expressed these ideas in a concrete form and has often been called 'the father of location theory'. One point is of interest to geographers: von Thünen's theoretical discussions clearly derive from direct observation – a comparison of the yields from two agricultural enterprises in contrasting localities with respect to market potential, and of the returns from his own lands situated 35 km from Rostock. The development of new types of relationships between urban and rural areas, strengthened by continuing urban growth, and the first signs of a major transformation of commercial activity, led to an economic differentiation in the rural areas that no conscientious observer could ignore. Von Thünen's personal records, as well as his theoretical speculations, allowed him to distinguish between 'natural locations' and 'economic locations'. He devised a system of peripheral zones around a central city reduced to a point, where the type of production varied according to distance from this point, reflecting the relative ease of marketing – that is, it varied according to the optimum location at which economic rent was maximised. This concept was based on concrete observations and is an approach to which geographers are perfectly amenable, an approach which many have applied once familiar with it.

Several of von Thünen's successors such as Roscher, researching intuitively and using descriptive methods similar to those of geography, analysed the factors of industrial location. Weber, however, radically modified the whole approach to industrial analysis when he projected 'pure economics into a spatial context' and derived 'a theory of location systems' (Ponsard, 1958). He conceived the idea of an abstract space, a 'new and empty zone', settled by successive human communities – consuming, organising, agricultural, industrial – occupying different locations and linked by a series of vertical relationships. Weber was the first 'to consider relating theories of regionalisation and of location' and to be aware of the 'possibility of the spatial association of productive activities' (Ponsard, 1955).

Several economists of the German school continued to pursue this approach: in addition to examining transportation costs, they introduced into their work on the definition of market areas the assumption of a differentiated physical space, generalised by regional scientists from Sweden in the form of isolines. Palander also made systematic use of this technique: he considered both climatic and legal factors, historical and technological developments that might have had a bearing on location. His theoretical study of various means of transport and the degree of competition between them could be illustrated by numerous examples from geographical publications, and the use of the now familiar isochrones was one of the issues he discussed.

After the First World War, however, changes in the economic climate soon became apparent: certain regions that had benefited from conditions in the nineteenth century entered a period of crisis; other areas, which had remained stagnant until this time, advanced in leaps and bounds and their new prosperity was marked by rapid urbanisation. The English and above all the Americans have been concerned with the modifications and upheavals affecting these clearly-defined areas, these 'regions' to use the most common term.

The work of Lösch is central to this research into economic regions. In his theory of location, he acknowledges an element of spatial choice and distinguishes between real and rational locations: the former involves 'the identification and explanation, either from an historical perspective or from consideration of "typical" behaviour, of the factors which effectively guide entrepreneurs in their choice of location', while the latter 'seeks to determine in the abstract the ideal location' (Ponsard, 1958). Economic regions demonstrate the interdependence between general and partial location: regions may be simple or form complex systems. The basis of the second type of analysis is an isomorphic surface, eliminating all variation that is non-economic in origin – a surface without geographical inequalities (that is, an undifferentiated physical space with uniform relief, a constant climate and of course no stream networks or valleys that would introduce morphological variation), without administrative boundaries, and with a common legal system and political equality – in fact a total economic uniformity in the distribution of raw materials and units of production, that is, of farms within a self-contained consumer society. These basic assumptions differ fundamentally from those of the geographer. The identification of market areas and hierarchical networks has nevertheless been a conspicuous feature of much of French economic geography, and research into urban networks and the hierarchy of urban functions directly reflects the spirit if not the assumptions of Lösch: these assumptions are more realistic and wide-ranging among geographers, and inevitably more narrowly economic among Lösch and his followers.

Heterogeneous and systematically ordered market areas precisely define an economic region, while the organisation of regions into networks constitutes the organic structure of space, through the interplay of purely economic factors on a simple uniform surface. However Lösch did introduce into his model variations resulting from both natural factors (accessibility, productive capacity) and human elements (behaviour of individuals or groups). These concepts were all based on the results of direct observation, together with 'a constant concern to relate the most abstract analyses to a concrete interpretation of reality' (Ponsard, 1958).

This is precisely where research by economists and geographers overlaps (if we insist on retaining specialist labels for basically common approaches!), since the theories of Lösch were directly influenced by those of Christaller. The two authors arrive at the same framework for spatial analysis. This excellent example is doubly interesting to the geographer: Christaller first examined empirically the existing urban network in southern Germany and its functional relationships; had he stopped at that point his monograph would have been forgotten, but he then formulated laws governing the nodality of space and the hierarchy of central places.

The validity and usefulness of his hypotheses were confirmed both by the theoretical reasoning of Lösch and by the model which he constructed. 'The Lösch model covers the full range of market areas but – as the author observes – all possible cases are not found in reality' (Ponsard, 1955). Geographers should take note of this lesson in methodology – and this conclusion: it is sufficient that a model should provide a theoretical framework representing the general case. There may be material that cannot be integrated into it, and a model may also define what does not yet exist or what we perhaps do not understand. Mineralogists working on the systematic arrangement of crystals, for example, left certain gaps in their system which were filled at a much later date, as with the discovery of hafnium.

Lösch therefore stands at the crossroads between geography and economics. There is now an increasing awareness of the need to consider spatial factors, particularly in relation to planning procedures, which often involves collaboration between the two groups of specialists. The work of Isard reflects this contact: his research is based on a wide range of case studies and together with formulating a basic location theory (1956), has been directed towards identifying the characteristics of an economic region (1960) – a complex functional organisation with a dominant focal point influencing a more or less extensive area.

This brief review of the development of spatial theory outside France clearly shows that economists cannot remain indifferent to physical space. Formulation of such theory has almost always coincided with the transformation of geographical space itself through technological innovations: Von Thünen was working at a time when the urban explosion following the Industrial Revolution had disrupted market conditions for peri-urban enterprises; Weber's research reflected a century of rapid industrial expansion. The region, no longer as a by-product of a general location theory but as a spatial division of interest in itself, has been the concern of economists in Britain and the United States for the past fifty years, ever since spatial variation in the rate of economic development and in the effect of economic changes became a political issue. This is particularly true today, as problems of regional planning in both capitalist and socialist countries become increasingly complex.

The work of these economists and of Christaller has also transformed certain areas of geography (Mikesell, 1969), but few Frenchmen have participated in this highly theoretical research. One or two volumes were published in France at the beginning of the twentieth century, concerned with the compilation of sociological data rather than its quantitative analysis, but only within the last twenty-five years have abstract concepts been widely discussed. Research has often remained highly specific, however, as in the excellent surveys carried out by the INSEE as an organisation or by its individual consultants (Klatzmann, 1956) and the study of the area of commercial influence of French towns, a monumental national investigation launched by Piatier (1970), the results of which are of great value to geographers. But for my purpose, which is not to present a panorama of French economic science, two types of research deserve special mention.

Firstly, there are economic studies devoted to 'regions', the boundaries of which have been determined for administrative purposes but which also have clearly defined physical limits. Gendarme examined the

Région du Nord (1954), and geographers should compare his approach with that in the *Atlas du Région du Nord*, completed at almost the same time (1959). Three features clearly emerge: the economist's concern with generalisation, his preoccupation with quantification and his total neglect of certain broader regional considerations. These three points are more revealing than long theoretical discussions on the differences of approach between the two disciplines.

Gendarme's concern with generalisation is brought out in the theme of his first chapter: he is not analysing the Région du Nord for its own sake but as a type example of an economic region, and he first discusses the broad geographical, economic and sociological definitions of a region. Geographers should read them for they demonstrate an unfamiliarity both with the recent work of geographers (the references cited are to Vidal de la Blache and Brunhes!) and with the objectives of their subject – 'geography is more descriptive, economics more explanatory'; 'geography treats only the visible aspects of economic problems, while the economist is equally interested in hidden or concealed phenomena'; 'we must reject the geographical definition of a region although it may contribute certain useful information, particularly to a study of agricultural production where climate and soil are significant factors'. On the other hand, there is a fair appreciation of one fundamental difference – the economist's concern with the region as an area defined by 'the problem under examination' and characterised by the interregional flow of population and capital. For Gendarme 'the region is an area forming a relatively homogeneous social and economic entity, the structure of which is dominated by a group of characteristic economic activities but which also has multiple relationships with other areas'. He is therefore stressing two essential features – homogeneity, emphasising a region's individuality in relation to other spatial units, and regional interdependence (Gendarme, 1954).

This concern with generalisation recurs in almost every calculation of an index or of summary statistics: in order to devise a classification that brings out the importance of the Région du Nord, or to specify criteria describing the growth of industrial enterprises, the author explains, discusses and interprets classical economic theory. Finally, this preoccupation is apparent in a third sphere – comparison with other regions and evaluation of the role of the state. Analysis of capital investment and of government legislation receives closer attention than the geographer would consider necessary, but often suffers from a lack of available data.

The second characteristic of Gendarme's work, quantification, is a result of this emphasis on generalisation. The author not only makes use of published statistics on population, manpower, output and profits – as does the geographer – but also tries to derive indices that will both define the region and reflect its pattern of social and economic growth. Numerous graphs and maps accompany the text and, in this respect, there is little to distinguish the work of economist and geographer except in the style of presentation. The introductory statement that the geographer would consider the Région du Nord as a closed space is outrageous, although it is true for example that he does not try to assess the value of interregional exchanges, a recent interest with certain economists.

Finally, we should note the near-disappearance in Gendarme's book of certain characteristic features of the Région du Nord. Agriculture is considered only through statistical allusion and urban life almost vanishes. This is surprising, to say the least, since the region is foremost in France in terms of the value of its agricultural production and second in the degree of urban development, whereas the author continually stresses the importance of industry as an element structuring regional space. The whole book has this emphasis, and both in plan and presentation it is in effect an analysis of theories of industrial location and development as applied to the Région du Nord, rather than a balanced study of the area. This illustrates a fundamental difference between geographers and economists: to the former the title of the book is misleading; to the latter it is simply the interpretation of reality according to its dominant economic characteristic. The Région du Nord is thus an area dominated by its industrial function and related trading activities.

In this context, it is useful to compare Gendarme's book with the thesis of Dézert, *La croissance industrielle et urbaine de la Porte d'Alsace* (1969), which has as an explanatory subtitle, 'a geographical essay on the evolution of a regional space as a function of industrial forces'. There are no generalisations, and no references to theories of regionalisation or to comparative models of industrial development. This fundamental contrast is absolutely typical of the different spirit in which the whole subject is approached.

Both volumes have considerable recourse to quantification – with the exception of the complex indices employed by Gendarme and ignored by Dézert – while the latter's graphs and maps are appropriately more elaborate. Dézert's space is far from being a closed area, however, and his investigation of the source of capital and trade flows in absolute terms (and not as a regional balance-sheet) shows greater insight. His description of the complex physical environment is also incomparably richer, not simply description for description's sake but specifically in relation to the dominant industry. The transformation of rural life, for example, is analysed not in terms of arid production and population figures, but with as great a depth and clarity as Gendarme displays in his own discussion of the history of industrial development in the Nord. Both volumes end with a discussion of the problems and potential for life in the region, but here too Gendarme concludes as a theoretician – 'the aim of this research has been to verify certain theoretical concepts'. He is disappointed on two counts however: 'the available evidence has forced me to modify certain assumptions'; 'the whole region will be subject to new and conflicting influences as wider European spatial structures evolve' (Gendarme, 1954). Dézert, on the other hand, is far more pragmatic: while demonstrating both the individuality (in terms of industrial influence) and the weaknesses of Alsace (the lack of an urban organisation), he concludes that it will become a growth pole within the European Economic Community (Dézert, 1969).

The necessity for collaboration between geographers and economists is obvious from reading these two volumes. One sees just how far an economist could draw on Dézert's work to illustrate the general theories the latter avoided and, conversely, all that a geographer could contribute to Gendarme's analysis which in certain respects is so shallow.

The second research interest, developed by contemporary economists, is that of 'regional balance'. The pioneer work in France, to which there have been a number of successors, was the economic description of Lorraine by Bauchet (1955). This balance is primarily economic but obviously includes two aspects where geographers could be of assistance: definition of the framework and the study of flows. The author specifies that he is considering 'the region as an assemblage of homogeneous structures', comprising both 'the firms operating in the region and the individuals consuming or working in the same region' (Bauchet, 1955). But in what region? Bauchet refers to a variety of publications, notably to the provocative articles by Cholley (1939–48), and states that 'the geographical limits to an economic region are determined by a synthesis of three related considerations: the basic requirements of individuals and of the major towns in a region, and the degree of self-sufficiency of that region' (ibid., 1955). While the first two criteria to some extent overlap and are fairly easy to delimit geographically, particularly the second, the third is a more abstract and difficult to define. The author is forced to adopt a solution which is not entirely satisfactory, either from a geographical or from an economic point of view, but which is of practical value since it has an administrative and statistical basis – an economic region comprising four départements. This example illustrates both the ideals of research workers and the constraints imposed by their data, as well as the role of history and the forces of inertia inherent in all structures. The provinces and more recently the départements have adopted approximately the same boundaries as those of the former Duchy of Lorraine, and we therefore remain prisoners of these limits to the extent that our research requires scientific data available only through official channels. But even if detailed statistics were published at the smallest scale, in France for example at the level of the commune, the geographer's work on regional delimitation would still be relevant. By regrouping the data he can suggest new definitions for spatial units and, as we shall see later, provide economists with the framework they require for specific projects. In addition, he must be concerned with the analysis of flows – their direction and significance if not their precise value – such as transportation networks, the circulation of men and of merchandise, trade and forces of agglomeration. The geographer's contribution to a study of this kind lies in his examination of both static and functional elements, whereas the treatment by economists will be far more theoretical and directed primarily towards the identification of economic systems, as in the recent book by Ponsard (1969).

An outstanding role in the geographer–economist relationship in France must be attributed to Perroux and his colleagues: while maintaining an economic standpoint they nevertheless make certain concessions to the geographer. 'Economic location comprises the determination of all structural coordinates, of all systems of relationships which have a bearing on the economic activity under consideration'. Firms choose locations where they are able to maximise profits, and their choice is constrained solely by transportation costs. Economic location therefore 'never coincides with geographical or geometric space as such: it is confronted only by prices and costs' (Perroux, 1954).

Economic space is primarily a space comprising multiple relationships: it may well be discontinuous and non-localised. The analysis of the great Anglo-Iranian oil company by the same author is an extreme example of this concept: it is space defined only by a plan of operation. In his various discussions (1950–64) Perroux in effect distinguishes three main types of economic space: firstly, *space as defined by a plan* – for a firm this operates within its own economic horizon (economic forecasts and targets within a specific area), while for a nation or a region it is the overall policy directives behind a particular project, once the viability of that project has been established (such as the examples discussed by Bauchet and Ponsard); secondly, *space as a field of forces* – the famous polarised space which is now so fashionable among geographers that it needs no further elaboration; thirdly, *space as a homogeneous aggregate* – over which there is considerable ambiguity. Perroux considers that each firm has a structure more or less similar to that of neighbouring enterprises. The firm belongs through the nature of its products or its price levels to a whole series of economic groupings. 'The structure of the firm is therefore more or less homogeneous with respect to each of these diverse environments.' Perroux is thus considering a structural homogeneity which may well bear no relationship to any localised geographical space. This is ambiguous, or rather a unique case which certain geographers have interpreted as a general situation. Homogeneity or heterogeneity of an environment is a highly relative concept, dependent upon the criteria employed in its definition. We may consider each nation as a unit 'comparable in its characteristic structural features to one or more neighbouring countries', or try to identify technological and regional groupings and the structures within each such group.

One can agree with Ponsard's general statement that Perroux treats space as 'the spatial projection of an essentially non-spatial framework'. He aims to de-localise economic activities, that is, 'to deal with abstract spaces defined in economic terms' (Ponsard, 1958).

These concepts had been widely accepted by French academics, although often in a superficial or tangential way. Boudeville and his colleagues, however, have taken them up again in several publications (1957, 1961, 1968) and made them at least partially more explicit. In so doing they have gone a long way towards finding common ground with geographers: there is now a firm foundation for rapprochement between the two disciplines and even Perroux recognises this. Although the location of an enterprise can be analysed purely in terms of economic criteria, more concrete factors should also receive attention in order 'to examine the firm within the framework of the more or less homogeneous structures that take account of its actual development: economic analysis, while preserving its own objectives, can also be seen as a transition towards geographical analysis' (Perroux, 1954). The author himself values this type of study, and discusses both the uniqueness of geography in relation to economics and also the potential interaction between the two sciences. He agrees with two statements by Gottman – that 'scientific economics must always aim to identify uniformities and is becoming increasingly quantified', and that 'geography should concentrate on the qualitative peculiarities of spatial facts'. These two distinct approaches should allow collaboration without rivalry or confrontation, 'provided that fields of study and methodologies are kept clearly separate'. Perroux continues: 'the geographer would therefore be concerned with space in so far as it is differentiated by the presence and activities of man, the economist in so far as it is organised to obtain the highest returns for the minimum cost and for the satisfaction of the needs of man' (ibid.).

In conclusion, one should re-emphasise the potential value of joint research, as long as each discipline preserves its own aims and objectives. For the economist the fundamental division of space is that of a market area, which expresses laws of production, consumption and exchange: it translates the constraints of geographical space into prices and costs. Let an economist conclude:

> an economic landscape is in no sense the horizon presented by nature, or as perceived by the eye of the geographer, the traveller or the painter. It is a rational landscape. It is the organisation of market areas as conceived by a logical mind. The theoretician of space . . . considers the concept of space and asks himself whether the real landscape conforms to his abstraction (Guitton, 1955).

Static space: the concept of homogeneity

Independent of its many other attributes, all geographical space has static components that are clearly visible on the Earth's surface. The subdivision of static space therefore requires only the regrouping of these components according to their inherent similarities: it consists on the one hand of identifying broad visual and physical boundaries, and on the other of grouping elements into homogeneous zones.

The concept of homogeneity deserves further consideration: nothing could appear simpler but in reality few things are more complex. Any definition is highly relative and varies primarily as a function of scale.

The lowlands of the Po valley are bounded to the west and north by the ranges of the Alps, just as the Massif Central borders on the Paris Basin. The most striking feature to emerge from a cursory glance at the 1:5 000 000 map is the strong similarity between plains and basins that lie adjacent to highland areas, whether these are mountain chains or plateaux. However, more detailed examination of the Massif Central reveals marked variation within this high compact region – tiers of semicircular plateaux in the northwest, raised blocks alternating with trenches and elongated ridges in the centre, rigid tablelands in the southwest, deeply and narrowly dissected into a long broken skeleton that falls away steeply to the east and south. There is not a uniform highland area but at least four distinct landscapes. At an even larger scale the basins of the Allier and Loire, extensions of the surrounding lowlands, are seen to penetrate close to the heart of the Massif and suggest further subdivisions. However, the geographer is forced by his intellectual integrity to concede that these two valley corridors have little in common – they are quite distinct in morphology, soil type, land use and settlement patterns. Their only similarity is their position adjacent to the mountain barrier which protects them, although this has highly significant consequences in terms of lower altitude, sheltered climate, routeways, high density of population and relatively rich farmland. One could argue that homogeneity through the juxtaposition of opposites is at least as valid as that defined by inherent similarity, and that criteria for the former can be derived mathematically.

As the scale is enlarged the number of criteria multiply, and it is only a small step to subdivide either of the above river valleys. The analysis by Derruau (1949) clearly showed that the Limagne comprises three districts: the hills and ridges of the west and south, the central marl plains – broad alluvial deposits of the game reserves along the Allier – and the Dore to the east. But if Brunet (1969) were to apply his classification of rural landscapes to this area, at least 100 homogeneous micro-units could legitimately be identified and described at a scale of 1:100 000. Above this scale, a search for homogeneity becomes pointless and the units already distinguished must be analysed in detail.

The concept of homogeneity therefore relates to the size of the area under consideration, to the complexity of the environment itself and to the degree of precision required. A vast sparsely-populated plateau or desert will almost certainly form a single unit at whatever scale it is analysed. Subdivisions of the interior plateau of Brazil are few and themselves cover vast areas, while a warped and dissected highland such as the Massif Central lends itself to the identification of numerous regional units.

There exists firstly, therefore, what might be termed *absolute* homogeneity. Where the physical environment is totally uniform, there is a single ubiquitous land use system and pattern of human settlement: certain purely agricultural areas of the central American plains conform to this definition.

There is secondly a *relative* homogeneity, that is, the even distribution of a single element irrespective of all other environmental variation. The extensive cultivation of cotton in the old American south dominates both the peripheral plains and low interior valleys of the southern Appalachians, while that of tobacco dominates the coastal plain, Piedmont and marginal valleys of the highlands. A deliberate choice must therefore be made to emphasise either the unifying or the divisive elements in a landscape. Provided that the selected variable is significant, that is, it is both visually distinctive and reacts with a complex of other closely interdependent factors, it is a perfectly legitimate choice in geographical terms. It corresponds to the uniform region described by American authors as 'an area of any size that approaches homogeneity in terms of the criteria by which it is defined: landforms, agriculture or soils' (Thompson, 1966).

To take two examples: the overall density of population in several areas may be the same, but consideration of density alone is inadequate since it may reflect widely divergent settlement patterns and occupations, for example, profitable farming in lowland areas or the clustering of families in a highland refuge and the exploitation of relatively infertile soils some distance away. On the other hand, the cultivation of cotton referred to above does correspond to a uniform climatic régime, to a population following a broadly similar life style and with homogeneous social and racial characteristics (descendents of the white planters – sometimes prosperous, sometimes 'poor whites' – and of the negroes formerly imported and used as slave labour). The economic problems and potential are almost identical throughout the cotton belt: there is obvious recession and parallel if not always equally efficient attempts at adjustment and transformation of the agricultural system. The only variation is in the size of the properties and of individual fields, which are larger on the lowlands and sometimes higher-yielding. Cotton cultivation is therefore one good indicator of relative homogeneity. The difference between these two situations can be summarised briefly: if an example of the first does exist somewhere – and one does not immediately spring to mind – it is a chance occurrence and geographically insignificant, although it may temporarily seem crucial during highly abstract analyses; in the second case, it is the clear expression of a set of interrelationships that are truly geographical.

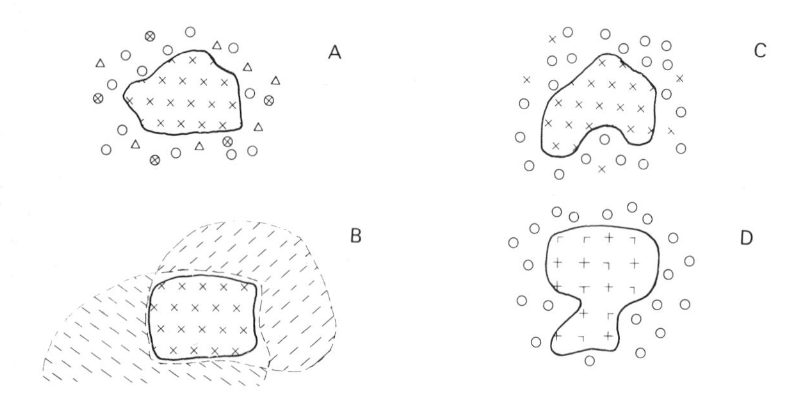

Fig. 3.1 Absolute and relative homogeneity
In examples A–D, the delimited groups each have at least one element in common and within-group variation is minimised. In group B, only identical and contiguous elements are included (absolute homogeneity). In example D, the complete element is a + which can be broken down into four units corresponding to the four arms of the cross: all elements have the lower vertical line in common (relative homogeneity)

Finally, there is *recurrent* homogeneity. We cannot refuse to describe an area as homogeneous when it comprises twenty-five separate valleys that are identical in morphology, in population distribution and in economic organisation: each valley is asymmetrical, with a long sunny and fertile *adret* and a cold uncultivated *ubac*. The one highland area is dissected twenty-five times by the same valley type. A monograph on each valley would highlight its own peculiar characteristics, while a monograph on the plateau would stress the recurrence of identical valley forms. The valleys and plateaux together constitute the homogeneous unit. Recurrent homogeneity would also be found in lowland farming areas, where larger villages were interspersed at 10 km intervals among a series of similar small hamlets. The problem is to assess at what point the typical pattern is so modified that the limits of the homogeneous area have been reached. This topic would need to be developed more fully and will be the subject of a separate monograph.

We should always aim to identify the most cohesive grouping, that is, where variation between the elements in a group is less than that between any two neighbouring groups (Berry, 1964; Brunet, 1969).

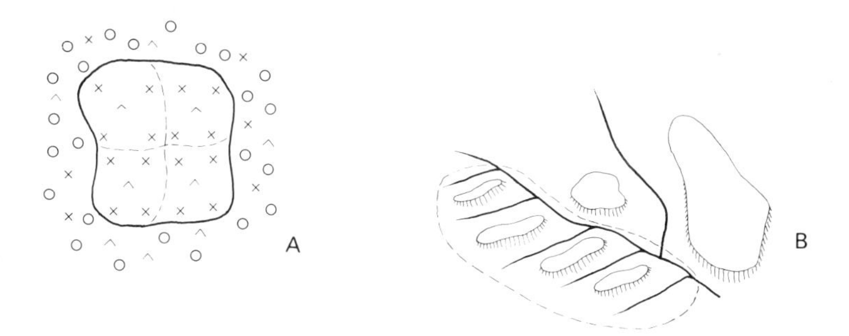

Fig. 3.2 Recurrent homogeneity
The delimited areas contain repeated elements which can be considered as a special type of homogeneity
A – distribution of hamlets and central villages
B – system of parrallel valleys

The Analysis of Space

Identification of an element

The geographer should follow two procedures when analysing space in relation to the characteristics outlined above: identification of the elements present and of their interdependent relationships.

The identification of characteristic elements is of course a function of the *a priori* considerations of the research worker – some will devote their attention exclusively to fisheries, others only to settlements, to the ordering of relief features, or to the integrated study of a single commune which could be in Brittany or Newfoundland – but generally-speaking, each specialist defines his own terms of reference and scale of operation.

We can illustrate this last point with reference to cartographic, graphic and other visual techniques, all of which are a geographer's basic tools. The initial element may be simple or itself comprise more detailed interrelated units. If we examine an aerial photograph, a tree is the simple element while a hedge consists of numerous trees and undergrowth: even a farm is a complex of buildings such as the farmhouse, sheds and stables; several farms together constitute a hamlet, and if we add a few residential and commercial units, and a few public buildings such as the mayor's office, a church and a school, it becomes a village. A field can be distinguished by its size, shape and colour, reflecting both the type of cultivation and the season in which the photograph was taken. From these observations, and with the aid of conventional symbols, it is possible to trace the location, shape and size of houses and villages – and also of fields, road networks and railway stations. But if we want to know the name of the owner of the fields or to analyse the complex land tenure and cultivation systems, we must examine a more detailed survey such as a cadastral plan. For a clear picture of the distribution of villages in a particular area, however, we need a smaller scale map on which individual houses will disappear, but distinct village units will remain . . . and so on . . .

The concept of scale is highly significant, and the appropriate scale will vary according to the nature of the basic unit. At 1:5000 or 1:10 000 we should work at village level or within a well-defined urban neighbourhood, and at 1:50 000 at the generalised village or urban district scale: at 1:500 000 the location of villages will be represented only by points and towns by minute circles . . . To examine a village, therefore we should either use a large-scale plan or preferably visit the area itself. However, for a general impression of the départements of Marne or Isère, or even of the whole Alpine chain, such a visit would allow study only of limited sections of the plateaux, valleys and mountains, and it would be impossible to draw a general map of the entire region from direct observation. An aerial view which allows us to survey extensive areas simultaneously is further illustration of this principle. Flying in a small aeroplane at a height of several hundred metres, we could reproduce detailed features of the landscape only over a limited area: at an altitude of 7000 or 8000 metres, however, we could take in Brittany, the Channel and the south coast of England at a single glance, or even the whole of Switzerland and the broad northern alpine foreland: Corsica and Sardinia would appear as tiny islands in a vast expanse of sea fringed by the coast of Italy, while photographs taken by satellite would reproduce whole continents on a single exposure.

We have so far considered only spatially extensive distributions, but we must obviously analyse all elements that can be represented at a given scale. For example, to study traffic flows within a town maps at 1:10 000 to 1:50 000 are adequate, according to the size of town and complexity of network, but to plot the actual elements in the circulation system, that is, vehicles and pedestrians, plans at 1:1000 or 1:2000 are necessary. To analyse the transportation network at the national level, a scale of 1:1 000 000 is sufficient.

Once the class and scale of the basic element have been determined, its characteristic features must be described and analysed. This is the purpose of a monograph – evaluation of an element's inherent structure and its position within a complex set of interrelationships. These relationships are basically of two types: vertical and horizontal – terms deliberately borrowed from industrial location theory. A cotton spinning mill is vertically integrated when it includes all production processes, from arrival of the raw cotton at the port to the sale of finished cloth on the Rue de Rivoli, and horizontally integrated when it is one plant in an industrial consortium that specialises in spinning. Exactly the same is true for any spatial unit.

Champagne-Mouton for example – with its evocative name – is located at the boundary of the crystalline massif of the Limousin and the limestone plateaux of Poitiers–Charente: its lands lie at the margin of both the grasslands and stockrearing of the Confolentais and the cereal polyculture of Ruffeçois. The market town itself has 971 inhabitants and a further 492 are dispersed over the neighbouring commune. It is the most populous and strongly nucleated centre on the plateaux surrounding the upper Charente valley. Several routeways converge at this point and it is therefore a true crossroads settlement, where densities of population are slightly higher than in nearby communes and are currently showing a marginal increase. The wide range of service functions in Champagne-Mouton introduces a certain diversity into an area where more than 60 per cent of the active population is employed in agriculture. It has all the trappings of a large market town and regional centre. Besides administrative buildings, there are adequate schools, hospitals and commercial facilities, as well as the offices of various agricultural organisations. The provision of these amenities, the town's dominance with respect to total population and occupation structure, and its location as a route centre all suggest that there should be polarisation effects. Detailed inquiries confirm this (Médevielle, 1970): people go to Champagne-Mouton to purchase consumer goods, to consult with the doctor, dentist or pharmacist, to visit government departments or a solicitor's office. The cooperative dairy collects milk from the immediate neighbourhood; there are also cooperative wheat silos and a shop selling fertilisers and machinery, while the Crédit Agricole invests in local farming activities. Children travel in to school from the surrounding area and both police and fire services cover several neighbouring communes.

The sphere of influence of Champagne-Mouton extends up to a radius of approximately 10 km round the town. For certain more specialist facilities, however, its inhabitants must in their turn look to Confolens or Ruffec, and for still higher order services to Limoges, Angoulême or Poitiers.

This brief survey illustrates two spatial characteristics. Firstly, the town of Champagne-Mouton is not a natural feature but has been implanted on the plateau for many centuries and is now one of its integral units. Secondly, it is surrounded by other units, other villages. Because of its size, its location and amenities, it performs central functions for the surrounding dependent villages, while it is itself related to larger neighbouring towns. It therefore belongs to a network of relationships, to a functional system. This is easy to demonstrate in a situation where human activity is clearly the dominant element.

Let us by way of contrast examine a physical unit – the margins of the Ile-de-France to the southwest of Reims. These famous asymmetrical scarplands today reflect both an initial set of physical conditions and the land use system introduced by man. There are firstly a series of 'vertical relationships'. A dipping series of hard conformable Tertiary limestones overlies the softer beds of clays and sands. Subsidence, weathering, chemical and fluvial erosion, acting jointly on this structure, have reduced the scarp to its present form while man has constructed villages and created a wine-growing district on the dip slopes. But the scarplands do not form an isolated area or a confined and enclosed space. They continue further to the north and south, and are themselves dominated to the west by the plateaux of the Ile-de-France and to the east by the chalk plateau of Champagne. These constitute the 'lateral relationships', again emphasised by human activity for while village lands do include the vineyards, they also extend to the scattered forest areas on the crests and to the cereal cultivation on the Champagne. These lateral physical relationships, reinforced by a land use system adapted to environmental variation and potential, form by their contiguity an indispensable link in a study of the scarp element. They also interlink with the system of vertical relationships, offering a series of evolutionary cross-sections. We may therefore suggest an outline classification of environmental characteristics:

Element A_2 Element B_2 Element C_2
Element A_3 Element B_3 Element C_3

- Structure
- Processes of Erosion
- Characteristics of the Physical Environment
  - relief
  - prevailing climate
  - soil type
- Exploitation by Man
  - cultivation
  - settlement
  - communications
  - trade

Element A_4 Element B_4 Element C_4

In a series of maps portraying the same information at the same scale, the lateral relationships of a given region will be shown on adjacent map sheets and they effectively define the properties of the region itself. It either has uniform attributes and therefore a homogeneous environment, or a series of distinctive subregions and a heterogeneous environment – atomistic if no groups of common traits can be identified, or structured if the boundaries of cohesive groups precisely subdivide the whole area.

> Whether one is arguing in favour of the uniqueness theory or whether one is arguing for regionalisation by way of classification and grouping procedures, it is necessary to define an individual or some basic unit of space to facilitate discussion. Two types of individual may be identified – the first by way of its space-time coordinates and the second by way of its properties. Geographic work has frequently confused the two and this has led to considerable confusion in the exposition of geographic problems and in methodological discussion in geography (Harvey, 1969).

All this may appear quite obvious: we should, however, recall the previous distinction between static and functional characteristics, as there are countless problems when these two concepts are confused – we shall return to this issue later.

Comparison of elements

I have deliberately refrained from using the classical term 'general geography' although it is essentially the approach in question – the selection of a single element and the systematic and extended analysis of this one theme.

Themes are as varied as are the issues relevant to geography: new topics are continually being explored and there is seemingly unending fragmentation of the subject. For a long time this was true particularly of physical geography, but gone are the days when Hartshorne could write that 90 per cent of the *Elements of Geography* (Finch and Trewartha, 1936) was devoted to developments relating to natural phenomena. That was the time when de Martonne and Demangeon were the sole representatives of general geography at the University of Paris. The number of chairs has since increased more than tenfold, from oceanography and climatology to rural, urban, political geography, not to mention specialists in particular zones – tropical, glacial . . .

There are several strands to this diversification. In certain general surveys a truly geographical feature is central to the research (as in studies of relief, settlement and land use), while in others the main element has an external origin (for example, in medicine or religious beliefs) and the geographer must evaluate according to his own methodology its cause and effect relationships with the physical environment. There is a wide gulf between the exact coincidence of geographical considerations and the selected element, and the point at which the geographer does no more than furnish a more or less detailed framework for its analysis. Dividing lines are often difficult to define and vary with individual research workers: it is useful, for example, to compare studies in medical geography by a doctor (May, 1958) and by a geographer (Prothero, 1965).

General geography therefore is the most seriously threatened with fragmentation, since its potential field of study appears infinite, and with adulteration, since it must constantly refer to work in other disciplines in order to achieve the objectives of its own analyses or to establish appropriate genetic relationships. A study of karst landforms, for example, relies on both morphometric and chemical techniques to identify a particular type of limestone, to analyse its degree of solubility and the composition of the water eroding it; it looks to climatology for an understanding of temperature variation and the relationship between infiltration and evaporation; it relies on botanists for data on the vegetation cover, which directly or indirectly affects the process of erosion; human geographers would explain the sequence of human occupancy. The controversy over the nature of the original vegetation of the Causses, and the disastrous overgrazing by flocks of sheep, is very familiar and is only one of the many examples where man has exercised a morphological or a climatological role. We must therefore be deeply conscious of the unifying spirit of geography, if we are not to be distracted by the potentially infinite ramifications of our discipline.

Conversely, a study in depth where structural analysis is pushed to its extreme limits, in terms of both cause and effect, is the most likely to stimulate mere acquisition of knowledge. Such knowledge may not be relevant to the final geographical synthesis but is indispensable to the general advancement of the discipline. Just as medical treatment by a family doctor improves as specialists make new discoveries in anatomy, bacteriology and serology, so an understanding of geographical relationships is deepened by the discoveries of morphologists, pedologists, demographers and economic geographers.

What empirical steps are taken by the various specialists? Despite the extreme diversity of research interests there are certain similarities in approach. The structural analysis of each element reveals its characteristics and internal dynamics (that is, the likelihood of evolution or transformation through time), its position and functions within a network of relationships. The elements then become the object of a series of operations which the geographer may perform simultaneously or in sequence.

The first step is to plot the distribution of analogous elements. Population geography is concerned with the distribution of men, climatology with the interaction between air masses, temperature, pressure and precipitation, giving rise to distinctive climatic regions. In both these examples we are dealing with dispersed phenomena, in the first case over the majority of land areas and in the second over the whole globe. Certain topics, however, concern more localised phenomena: urban geography and glacial or granitic geomorphology deal with large but discontinuous units. We can therefore examine either single elements with a world-wide distribution or limited areas including some but not all of the same features.

The second step is obviously to compare these elements. Urban geography, for example, attempts to define an urban area and analyses the distribution of towns. It takes into account differences in size, in architectural style, in functional structures and mechanisms and then suggests a typology, or rather typologies, of urban areas with respect to individual criteria. Grouping of these criteria may permit the formulation of a general classification. The same procedures are followed in all branches of geography, although the typologies are more or less clear-cut and elaborate according to the particular topic. Several classifications of climatic types have been put forward, for example, some of which overlap while others are complementary.

Description of selected elements and their distribution, followed by comparison and classification, whether based on a single criterion (for towns on total population, in geomorphology on rock type) or in relation to time and space coordinates, whether complex or universal, are the series of operations within what is traditionally called 'general geography'.

What is the role of space in these considerations? Space is of course important but is not crucial, since it is essentially considered in a discontinuous sense. Firstly there is discontinuity within complexity: a single element is selected and related environmental factors are then explored, either within the narrow limits of a particular theme (geomorphology or climatology, for example) or essentially to set this theme in a wider perspective. Population geography therefore does not focus on relief, climate, soils or communication networks *per se*, but considers them in relation to the number of human beings, to their movements, needs and economic activities (Beaujeu-Garnier, 1956–58, 1969). Secondly, there is discontinuity within space itself, which is not invariably the case but occurs when an element is highly sporadic: urban geography must ultimately relate to terrestrial space, but this space may be vast and contiguous, as in a megalopolis, or form a point pattern, as when towns are separated by farmland or mountain ranges, or include a group of islands, so that the total urban area is far greater than its actual land surface. Finally, there are branches of geography where spatial considerations are reduced to precise geometrical units, such as lines of communication, or are appreciable only through indirect indices, as with the geography of capital flows.

This sequence of steps in general geography, applicable to highly diverse phenomena, is a self-contained set of procedures: we are not concerned with the objectives of geography but simply with an empirical methodology. We are no longer considering general man–environment relationships but only partial interaction: man may be ousted by a whole range of physical considerations and the environment may not feature at all in certain studies directed towards economic problems.

Although a discontinuous spatial approach is indispensable to overall progress in research – and there are many examples to support this – it nevertheless poses a threat to the future of geography and to any defence of its uniqueness.

Grouping of elements

An alternative approach is to analyse the complete vertical structure of a number of elements and their interaction in continuous physical space – an approach normally and inappropriately termed 'regional geography'.

Why should there be a separate 'regional geography'? From discussions so far on the nature and methodology of geography, it is obvious that most complex phenomena have a specific location, a limited area in which they have been wholly or partially implanted. The expression 'regional geography' is an empty shell which individual geographers attempt to fill, at the same time pursuing lively arguments with specialists from other disciplines such as economics over this simple word 'region', which is credited with an extraordinary range of meanings.

One initial assumption must be clearly stated: geographical methods can be applied within any areal framework (Gourou, 1970). Geographers must analyse a unit of physical space, and investigate the associations and linkages between all the elements present in that space. This is 'special geography' – or simply geography – although it is now common to use the phrase 'regional geography' to distinguish it from systematic branches such as physical, human, rural or industrial geography. This misuse of language is the source of a triple misunderstanding.

Firstly, certain linkages have a pronounced spatial expression resulting from industrial developments, from human activities of various kinds or from the system of land use and rural way of life, and there is therefore a geography of industrial, urban, rural and administrative areas. But should one separate the study of these areas from straight industrial, urban and rural geography? This would lead, for example, to an industrial geography severed from its spatial roots and therefore more correctly called 'industrial economics'. Such separation could only be to the detriment of geography as a discipline in its own right.

The suppliers and customers of a large industrial enterprise may be located in the four corners of the globe, but the firm itself still has a precise location. It attracts a labour force, and stimulates both transportation flows and an increase in purchasing power within a given area: its own means of communication, the nature of its market and potential for expansion may also be directly related to its location. Précheur (1969) claimed that 'the majority of industries today are no longer constrained by time and space' and emphasised the role of capital flows. However, this is perhaps lyricism or frustration on the part of an author who several years previously had clearly demonstrated the spatial impact of the iron-smelting industry of Lorraine, and who had deplored the fact that problems of documentation and the broad scope of the project allowed him no time to study 'the men, the labour force and historical developments' (Précheur, 1959). Determinism in general should not be confused with concrete situations: developing industrial centres may be free of traditionally acknowledged constraints such as proximity to sources of raw materials and the importance of a particular means of transport, but they still have a precise location and a spatial impact, however short-lived.

Secondly, general and regional geography are often seen as a dualism, but does this imply that there is no general regional geography as there is general physical geography, general human geography and general geography of transport? Does it imply that although relief types, populations, urban centres and rural landscapes are comparable, this is not true of various spatial divisions? This is an obvious absurdity.

One might argue that the recurrence of an individual element in numerous areas is according to mathematical laws, more likely than that of a combination of elements. As local features become more complex, their exact reproduction is increasingly unlikely. However, regional analysis in depth frequently shows that uniqueness is more apparent than real, or rather that fundamentally similar processes are in operation behind the 'unique' surface phenomena: this requires a distinction between essential and subsidiary characteristics. It has already proved possible to compare traditional rural structures from widely-separated areas, while in both France and the United States, certain analyses in urban geography have ignored administrative boundaries and have included problems of the urban fringe in their comparison of distinctive urban spatial structures.

The transition is straightforward: it is simply a case of applying the methodology of general geography, that is, of comparing not only simple structures but also complex combinations of elements, and of stressing their underlying similarities. Cholley's claim that geographical facts are unique must be modified: it may be a useful assumption in intuitive analysis but is dangerous in any later attempt to draw a comparative synthesis, which should be the geographer's ultimate aim.

Brunet (1968) has put this very succinctly:

> Historians like regional geographers must of course study individual personalities and isolated incidents, but only if armed with a wide background knowledge and a clear under-

> standing of evolutionary processes. On the other hand, comparison of individual events is itself a means of identifying processes and relationships, which then help to explain other 'unique' situations. The recognition of types and the comparison of complex groups within each type is the fundamental task of the geographer . . . even regions can be classed in this way, and the comparison of similar regions is a particularly rewarding if exceptionally difficult task.

Finally, this so-called regional geography has a third failing: geographers have become hypnotised by the idea of the region, the problems of the region, the concept of the region. An extensive literature has developed from this obsession and given rise to endless disputes. Regional geography has alternately been considered a general lumber-room and the quintessence of geography. Provided that they relate to one area (however defined – a nation-state, province, county or département), twenty successive chapters on topics ranging from geological history to sociology and political life would constitute a work meriting the label 'regional'. It is apparently sufficient to name an area in the title (industry in Lorraine, the population of New England, crab-fishing in Kamchatka) for a wide range of themes to be included under 'regional geography'. This is legitimate, however, only if they are considered as part of a system of relationships operating within the area. In Lorraine, for example, mere enumeration of the various local industries, with a summary of production figures and technological data, is inadequate: we must bring out the contrast between the two industrial 'hearths' – the large agglomerations of heavy industry in the north and the residual or modified artisan activities in the south.

We must therefore renounce this narrow view of regional geography, which amounts to equating it with one of the many other systematic branches – almost as a rival. We should restore its true status as the science of concrete spatial analysis. Outstanding geographers such as Vidal de la Blache claimed from the first that 'regional synthesis is the ultimate achievement of a geographer, where he is most fully himself', while Brunhes wrote that 'regional geography, in its broadest and most general sense, must be the crowning synthesis and not merely the beginning of analytical research in geography'.

This type of analysis must be dynamic and as flexible as the complexity it seeks to penetrate and understand. It is not the monopoly of any one group of geographers but the objective and ambition of all, the unifying trait which, despite their highly diverse and esoteric research interests, distinguishes all geographers from specialists in other disciplines. Thus 'regional study . . . will remain what it was always destined to be . . . one of the supreme sciences of man . . . and an end in itself' (Le Lannou, 1949).

The place of regional geography as 'the citadel of our whole subject' has been widely acknowledged for many years and its definition by numerous scholars has remained basically the same. In Germany, Hettner claimed that 'those who are not concerned with understanding the countryside [regional geography] always run the risk of neglecting the essentials of the subject, and cannot be true geographers' (1919). This was recently stated even more explicitly by Gilbert: 'geography is the art of recognising, describing and interpreting the personalities of regions' (1960).

Regional study – or the study of combinations of elements forming a geographical unit – must therefore be given priority and constitute both the basis and the ultimate goal of our research. But how precisely does one delimit and analyse these groupings, and what is meant by a 'geographical unit'? From discussion so far, it is clear that they are defined according to the characteristic relationships between elements in a limited area, either as homogeneous elements within a static framework, or as specific functional relationships operating with a given degree of intensity and continuity.

Static elements and functional linkages

It should be stated quite categorically that in every branch of the subject there is a fundamental distinction between static and functional geography. The traditional dichotomy between general and regional geography is illusory since it is the variation in initial research themes which is significant. These may be either static and visible on the Earth's surface, or the expression of a movement, of a relationship which is at first invisible and must gradually be unveiled.

To compare two unlikely situations: the general geography of glacial areas is concerned with the agents, processes and effects of erosion, and with the distribution of glacial features. It then attempts to formulate general laws governing glacial erosion and the evolution of relief forms, but considers only isolated spatial units that have no functional linkages. In contrast, the so-called geography of international trade examines demand and overproduction, organisation and methods, and flows between complementary areas. It rarely ventures into theories of exchange, which are the province of the economist. This situation is one of disequilibrium and flux, in which the geographer concentrates essentially on tangible evidence, that is, on exchange flows. In the first case, however, the basic theme is a relief form inscribed on the Earth's surface.

It could be argued that this is no more than a distinction between physical and economic geography – but both are traditionally classified within the one category of general geography. There are also examples where the situation is almost completely reversed, such as the morphological study of static urban spaces or of stream networks as functional units.

How on the other hand should one classify types of industrial region or types of urban region? As general geography since it is a question of 'types', or as regional geography since 'regions' are involved? These inconsistencies deserve more serious attention.

Conversely, a glacial valley can be studied both as a unit in itself and also as a routeway and an environment for settlement; a forest can be considered as the joint expression of soil and climatic factors, the limit of cultivable farmland, a source of timber and of cool air, a leisure area or an obstacle to communications; a town may be analysed as a static space or as an inhabited area and growth pole; a commercial district is a list of enterprises or a focus of commercial activity; rural areas constitute regions of agricultural production or a pattern of dispersed villages, silos and cooperatives, dependent on external organisations. There is always a duality of relationships which must be acknowledged in the final stages of geographical research. Neither presents an identical image of the world in which we live, but both are essential to its analysis and therefore to any comprehensive explanation.

The framework for a full analysis of these 'geographical units' must be revised. Geographers in many countries have been attempting to do this for a number of years, but often without adequate consideration of the distinction between static and functional approaches outlined above.

First, however, we must conclude our discussion of the word 'region'.

The static and the functional in geographical space

Parallel with an analysis of the different approach to spatial issues by geographers and economists, we must examine the dualisms within geography itself. Two distinct procedures should be followed for an effective analysis of our complex field of study, taking into account the universality of phenomena and the need for systematic study, and also preserving the goal of active participation in land management planning.

Personal survey of an area and examination of basic documents provide a static impression of particular elements or spatial units. The relative uniformity of the Beauce is immediately apparent in its relief type and surface clay deposits, the proportion of land under cereal cultivation, the absence of hedges and the pattern of village settlements. On the other hand, even a cursory survey of the area within a 30 km radius of Lyon would reveal striking contrasts between the green, undulating and dissected hills of the Monts du Lyonnais, the many ponds and lakes of the Dombes lowlands, the plateaux and hills giving way in the southeast to the arid Chambarin and the pre-Alps, and the urban complex of the second metropolitan area of France.

An initial survey of static elements will therefore highlight the uniformity of the Beauce and the fragmentation of the Lyonnais. But the geographer must go further: he must also examine the various forces operating within each space. Men inhabit the Beauce as well as the plains and hills of the Lyonnais. Certain of their activities take place within the region itself – in their fields, villages and small towns – while others involve some of the population in more or less distant travel. Human agglomerations of whatever size, and however varied their physical environment, are not isolated: they are linked by communication networks and exchange flows. A common administrative structure, or a hierarchical organisation such as that of a commercial enterprise, also promotes contact between settlements. Research into functional structures, that is, a dynamic study of the linkages between static elements, is equally important to the geographer.

One very familiar image will illustrate this point. We can take either

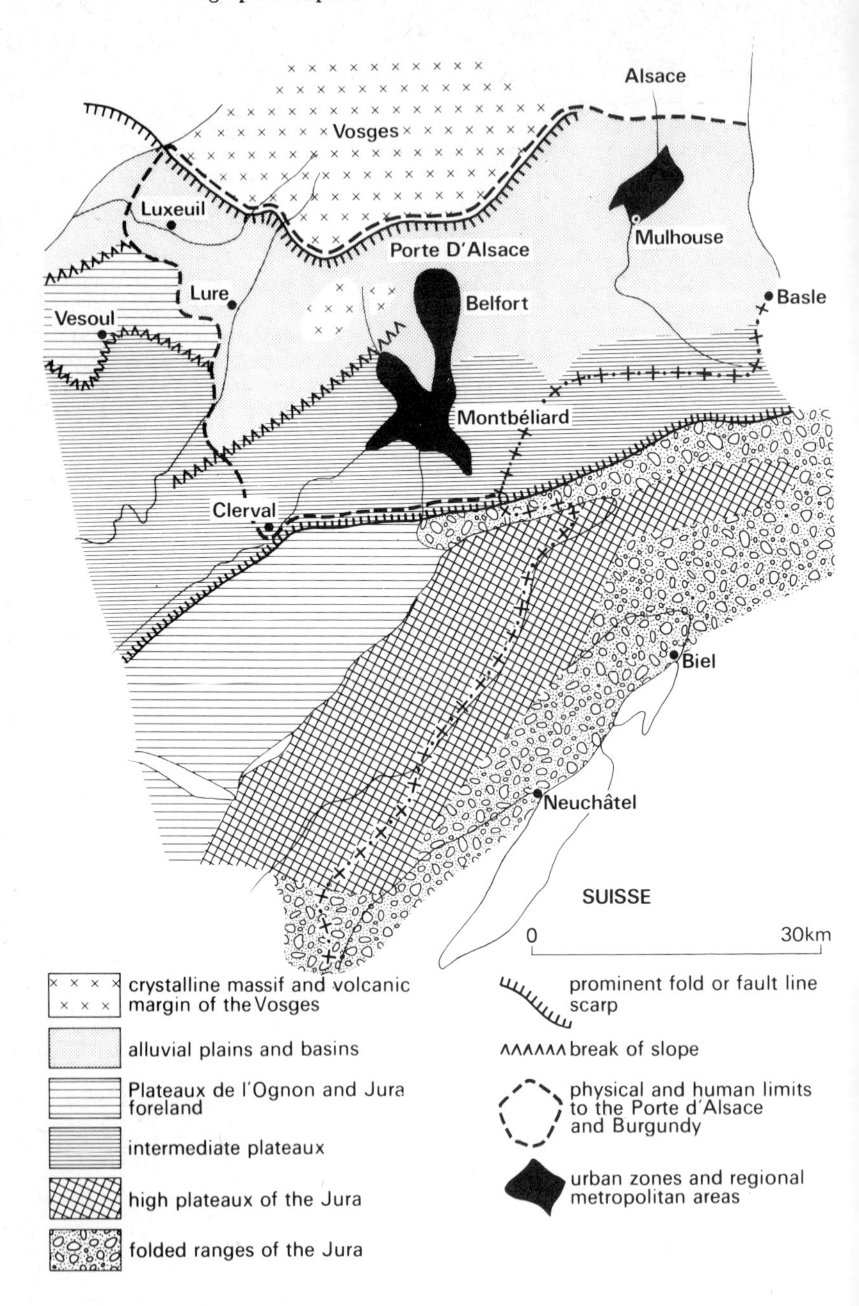

Fig. 3.3 The natural region of the Porte d'Alsace
An example of the static division of space (after Dézert)

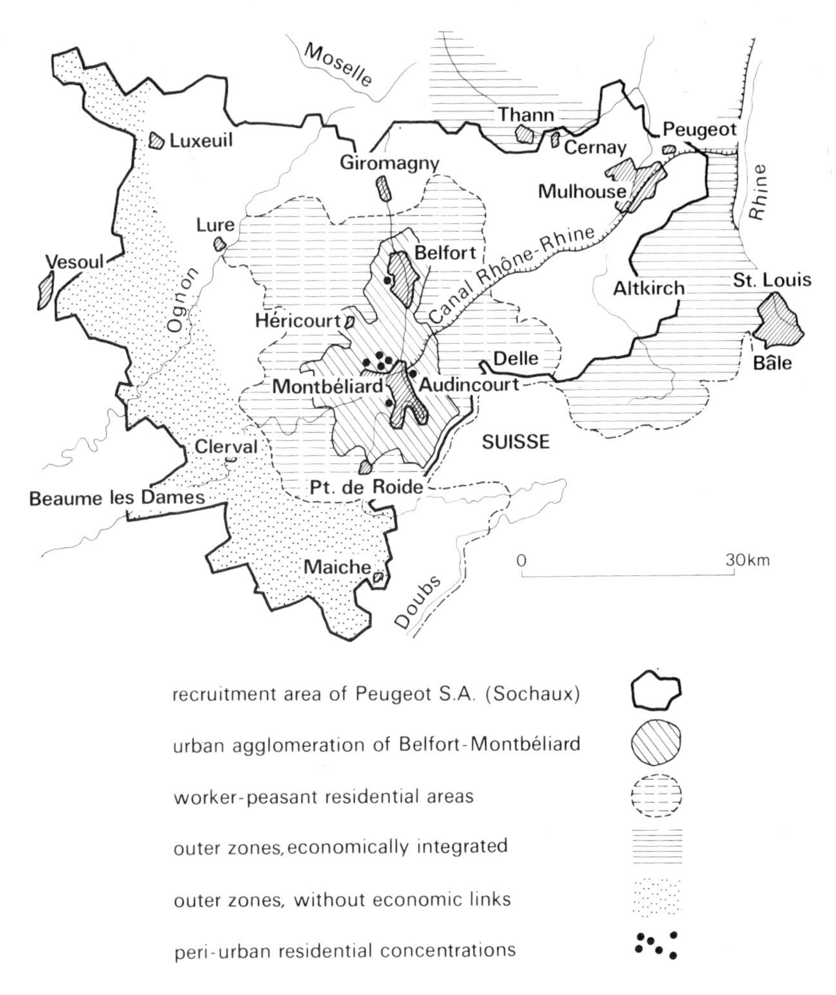

Fig. 3.4 Sphere of influence of Peugeot S.A.
Functional division of the area shown on Fig. 3.3 on the basis of urban and industrial organisation (after Dézert)

a snapshot or a cine-film of a given situation and thus obtain two separate impressions of the one reality. The former is fixed: it contains the potential for action but although forces may be apparent – as in a snap-shot of a boxing match where the boxer's gloves are thrust towards his opponent – they are not yet operative. A photograph sets the scene and introduces the characters, which has a certain objective interest, but once a film is unwound everything comes to life: fixed postures are relaxed, people move; there is suddenly a new angle, a new way of interpreting gestures and personalities. The tableau becomes alive.

How satisfactory therefore is a still photograph? A traditional panorama of a landscape may be quite adequate, since it gives a clear impression of the shape of the mountains and the nature of the vegetation cover. However, even the most imaginative photograph of a crowd in a street or market, or of a motorway jammed with cars, leads only to a series of unanswered questions – where is the source of these mobile elements? what are they doing? what will they be doing in the next few minutes? A cine-film, on the other hand, allows us to trace their activities or part of their route for at least a few minutes: a train may come down a deserted railway track, a light appear in an empty shop window.

Two perspectives are therefore superimposed, and both confusion over geographical classification and fruitless discussion on the meaning of the word region could have been avoided if more geographers had clearly understood this. 'A spatial system comprises places, the attributes of these places, and the interactions among them' (Berry, 1968). We must describe spatial attributes in order to identify homogeneous units: a functional definition of these units then allows us to delimit interrelated groups or areas. The first perspective, which is equivalent to regionalisation in economics, leads to a spatial mosaic in which all the units are internally homogeneous but vary enormously in size and characteristics. The second perspective, described by economists as 'polarisation', leads to a series of functional spaces in which interrelated flows of varying type, strength and direction are traced over the basic mosaic: when these flows converge on a number of central points, the pattern is one of a spider's web.

It is extremely important that these geographical approximations to the terms defined by Perroux should be used only in the light of our previous discussions. It is not through a misplaced desire for originality, however, that I suggest adoption of static and functional rather than of homogeneous and polarised. All spaces have static and functional aspects: there are static elements in both homogeneous and heterogeneous areas, and it is inadequate in a geographical discussion to consider only the former. Functional areas may not be related to central places in human environments that have a weak system of spatial structures (under-developed areas) or are over-urbanised (an amorphous agglomeration of mining centres), and functional relationships may therefore appear as a series of sliding and rumbling movements, a true Brownian motion, rather than as a regular pattern of focal points.

In summarising this wide range of concepts, it is tempting to state that there are two spatial perspectives – static and functional – and two criteria for the grouping of spatial units – homogeneity and polarisation.

None is of greater significance than any of the others and no geographer should give priority to static rather than functional considerations or vice versa. Although they may not have an equal impact on a given area, both are important in relation to its physiognomy as well as to its basic properties: certain areas are characterised by static elements while others are essentially defined by a series of functional linkages. In some cases, the linkages may so overshadow the static units that these must be re-grouped or further subdivided: in others, static elements are so dominant that they severely constrain the operation of linkages, or reduce them to highly localised epiphenomena within vast areas. Any combination of these two situations is possible but is not a random occurrence: it reflects the momentary equilibrium between two orders of facts – the strength and attributes of the underlying elements and the impact of the activity of human societies.

But why have I selected the specific terms 'static' and 'functional'? The choice of static was made only after examination of all its ramifications. In mechanics, static refers to the 'study of systems of physical elements subjected to forces which create no movement'. I have previously defined geographical space as 'an aggregate of elemental units . . . linked through the interplay of forces'. The concept of inertia must therefore be rejected from the outset, since each element in a geographical matrix has an inherent potential for dynamic transformation and is linked with other elements in the same matrix.

The appearance of a valley is continually undergoing change, but at any one moment it reflects a balance between systems of opposing forces. At time x these forces are equal and no movement is created: at time $x + n$, the point at which the film was unwound in our previous example, particles begin to slide, drops of water infiltrate, soils are broken up and the equilibrium is destroyed. Similarly, the employees of a shop are expected to be in that shop at a specified time and their activity is concentrated in a limited area. A bell marks the end of the working day and all the assistants then leave the premises, each motivated by the desire to return home. In successive time periods, forces are therefore released which provoke a rupture followed by a new equilibrium, and it is clear from these two examples that the process is irreversible. Forces already exist within given elements and operate in a variety of ways through time to produce a sequence of equilibria. The geographer must select his own time x, a limited period during which the situation is stable and he can then attempt to analyse its complexity.

This approach might be considered unsatisfactory since there is an arbitrary cut-off point between static and functional elements. It reflects a fundamental aspect of research however – the necessity to define separate stages along a continuum. In the film of a race, for example, the fixed point at the winning-post determines the victor, although sometimes by only a few centimetres. In a study of commuting, it is crucial to record each person or vehicle both at the destination on which they converge and at the start of the journey, at the numerous points of origin. In urban residential areas we consider such criteria as density, poverty level, consumer demands and significant behavioural features: it is equally essential to consider individuals from these areas at their place of work, in a different locality and with different densities and demands. The presence of the same people in two places poses specific problems which are not always complementary. This is analogous to the view taken by Berry (1964), when he suggested recording information on a given region within a three-dimensional framework – columns of place names, rows of attributes of these places, and consideration of the same rows and columns at appropriate cross-sections through time. The only way to comprehend certain comparative and evolutionary processes is to analyse a sequence of still pictures of the whole complex. The definition of the framework determines the basis for planning decisions in a particular environment. It is dangerous to launch such grandiose schemes as the development of a growth pole, for example, without an understanding of the basic mosaic, since each component may behave according to its own laws.

The term static has long been in common usage, particularly by economists, as the traditional antithesis of dynamic. Why therefore have I not suggested preserving this classical dichotomy? I prefer the term functional to dynamic for two reasons: the first is precisely because dynamic does imply the opposite of static, whereas it is important to convey the idea of a dualism, of superimposition, or more accurately of complementarity. Secondly, dynamic also implies an undefined potential for future development: 'the dynamic, the science of ignoring time' (Guitton, 1955). Geographers, however, are concerned with the processes which bring about that development, with the means of realising the potential in each element – 'geographical environments are the battlefield for functional relationships' (Baulig, 1959). These instantaneous relationships are interrelated with and parallel to the static elements, and are not a consequence of them. The geographer's definition of functional therefore approximates to the mathematical expression 'functional calculus'.

In the light of the discussion so far, it is obviously impossible to make a distinction between criteria valid for an analysis of the static environment and for a study of functional relationships. An examination of the whole range of relevant geographical factors would be necessary in each case. We should therefore set ourselves the more modest task of attempting to identify 'characteristic reagents'. Two comments are useful at this point. The static environment is explained primarily as a result of direct observation, while functional areas require an understanding not only of visible flows but also of a series of intangible and highly complex interrelationships. Observation of two static environments, or of the same environment at two successive time periods, often leads to the discovery of a functional linkage and is the key to analysing the mechanisms involved.

For example, to fly low over the Paris conurbation is a splendid way to develop an awareness of the visually distinctive characteristics of a large urban agglomeration: the cramped old houses and narrow streets of the city centre, the irregular pattern of high-rise properties in the inner peripheral zone, the star-shaped or neatly zigzagging suburbs encroaching on the remnants of forests and fields, and finally the tranquility of the countryside where isolated villages still remain. Dividing or linking all these areas are the peripheral ring roads and the great radial highways. The structure of the whole city, and also of each quarter, is defined and emphasised by the skeleton of the road network. Long avenues of high-class residences or neat squares complete with their statues break the monotony of the drab buildings. The leafy canopy of the open spaces, the blue-green of the rivers, and the emerald green of swimming pools and flooded quarries stand out above the greyness of the houses and factories and the green patchwork of suburban gardens. There is no substitute for the aerial view of a town: more vividly than any lengthy description it reveals to the geographer its fortunes and mysteries. An old quarter may lie immediately adjacent to a new housing estate: a compact and chaotic French suburb bears no resemblance to its German, English, American or Australian counterpart – monotonous and open-plan. In Anglo-Saxon countries suburban developments of the 1940s and 1960s can easily be distinguished, the winding crescents contrasting with the grid-iron pattern of the older estates.

At ground level, analysis of static features comprises observation of types of property in both residential and industrial districts, of the level of commercial activity, and of the density and total number, employment, demographic and social characteristics of the inhabitants. Static description is therefore strongly orientated towards 'the container and its contents' (Perroux, 1955).

Towards half past five on a Thursday evening, there is a terrible congestion of pedestrians and vehicles at the Place de l'Opéra in Paris, while at the same time on a Sunday it is almost deserted. Conversely, in Grosrouvre or Courpalay, or in any one of the numerous outlying villages, the shutters of most houses are firmly closed throughout the week but overlook carefully tended lawns and flower beds. On Saturday and Sunday, however, cars bearing familiar Parisian registration numbers – 75, 78, 92 – are parked outside the gates: windows are flung open and lawnmowers are in action. Observation of static elements at regular intervals may therefore reveal considerable variation in their outward appearance. A comparison of the permanent and weekend garaging of the cars would demonstrate the underlying functional relationship: some indication of this would also be gained by standing on a bridge overlooking the multicoloured caterpillar movement of cars along the great axial roads leading from the capital. The basic elements in this visual representation of functional linkages between various sectors of geographical space are therefore the units in the circulatory system.

Having first examined subdivisions of the mosaic space determined by static characteristics, we must then seek to identify patterns of circulation in its broadest sense, that is, circulation of men, capital, ideas, vehicles and products: this applies at all levels and in all fields of study. Differences in the organisation of geographical space, and therefore in the possibility of recognising spatial structures, result from the various combinations of these patterns.

The myth of the region

Few terms can be as imprecise as the word region. It means all things to all men, and is used in both colloquial language and in the technical vocabulary of economists and administrators, in both geographical descriptions and tourist brochures. Sometimes it refers to a vast agglomeration of areas, sometimes to a small district. We also use such expressions as 'the temperate regions of the world', 'the Great Lakes region' of the United States, and 'the Paris region'. A Parisian will even state that he spent his vacation in the Pontoise region, referring to an area of a few square kilometres in the vicinity of that town. The word is used indiscriminately to describe an area with uniform physical characteristics, shown on a topographic map in a single colour (brown for mountains, green for plains, and so on), such as the Massif Central, the lowlands of Scotland, the mountains and plateaux of western North America, and also to areas of highly diverse relief. The region of Midi-Pyrénées in southern France, for example, comprises the edge of the Massif Central, a trough of lowlands and the central ridges of the high Pyrenees, while the 'Deep South' of the United States covers the whole southeast of that country.

What is a region?

To demonstrate the paradox of the situation, we could site the wide-ranging definitions of a region to emerge from a survey conducted in France a few years ago.* For 56 per cent of the respondents, the feeling of belonging to a region derived from the fact that their parents lived there and that they themselves were born there; 46 per cent mentioned similarities of temperament and 44 per cent a pleasant climate; for 28 per cent the landscape itself was significant, for 25 per cent the region's past history, for 22 per cent the dominant industrial and commercial activities and for 21 per cent the agricultural system; others evoked common interests and traditional folklore, but only 11 per cent mentioned the influence of the nearest town! These results would be somewhat depressing if it were not for the fact that people from the Mediterranean areas emphasised climate and temperament, northerners their industrial and commercial activities, inhabitants of the Beauce their agricultural system and of Rhône–Alpes the influence of a nearby large town – which is good reason for believing in sound common sense as an invaluable guide to scientific and abstract thinking! Similar responses probably led Labasse to state pragmatically that 'for its inhabitants . . . the region . . . is a way of life' (1960).

A survey by Piveteau (1969) of regional awareness in Switzerland is also instructive. It was conducted among students at the University of Fribourg and showed that this awareness varied according to their 'sex and course of study, and in particular to the location and size of their home town and their degree of residential mobility'. In general, 'regional consciousness applied to relatively limited areas': for one-quarter of the students it did not extend beyond their home town, for 11 per cent it was the canton district and for 10 per cent the canton itself (roughly equivalent in size to a French département).

The same discrepancy between popular sentiment and academic concepts would be apparent from a survey of many rural areas. The Morvan, for example, is a small and clearly-defined unit in terms of both its geology and geography, its terrain and human activity. Yet a farmer from Saint-Aignan in the centre of the plateau, whose fields are fragmented to form a minute patchwork on the residual Liassic soils, will claim that his are the fertile lands and point to the south of the Morvan 'with its ill-wind and hostile inhabitants'.

The survey by the IFOP demonstrated a similar uncertainty over the

---

* In Guichard, O. *Aménager la France*, Paris, Geneva, 1965, 197–8. This is an extract from a wider investigation undertaken by the IFOP and published in full in the first issue of the review *Sondages* for 1965. The total exceeds 100 per cent, since several of the questions elicited multiple responses.

size and boundaries of a region as over the concept of a region itself. In answer to the question 'to which region do you belong?', several respondents referred to an extensive physical complex (the Paris Basin, the Alps, . . .) and others to general cardinal points (north, east, south, . . .): there were frequent references to the départements, and also to the former provinces (Champagne, Lorraine, Alsace-Lorraine, Bourgogne, Poitou, . . .), to the traditional 'pays' (Brie, Beauce, Cévennes, Camargue, . . .) and finally to small urban or industrial districts (the Bassin de Longwy, the region of Forbach or Nantes, . . .). This is evidence of the extreme eclecticism of personal opinion at the present time.

A research worker hoping to test popular opinion would be bewildered by the cloud of ambiguity surrounding anything to do with the word region, apparent in its varied usage and contradictory definitions. Popular usage could not be more vague and subjective, yet the same word has been endowed by certain specialists with all the virtues of a precise and detailed definition. It is hardly surprising that there is so much dispute over this unfortunate word.

Definitions of a region by geographers and economists in particular, but also by sociologists, historians, administrators and politicians from all over the world, would undoubtedly run to several volumes. Odum and Moore (1938) collected over forty such definitions, a few of which merit quotation in order to illustrate the wide range of viewpoints: 'an area with homogeneous physical characteristics' (Joerg); 'in between the continent and the village lies an area sometimes larger sometimes smaller than the state – it is the human region' (Mumford); 'regions are distinct units, each of which exhibits both physical and cultural differences from its neighbours' (Renner); 'an area within which the combination of environmental and demographic factors has created a homogeneity of economic and social structures' (Woofter). The list could be extended but the conclusion would be the same: whatever definition is adopted, opinion is unanimous on one point – the region is a spatial unit which is clearly distinct from neighbouring spaces. But in what ways is it distinct? Here viewpoints again begin to diverge and Meynier (1969) has shown just how wide this divergence is: let us resume the debate.

Types of region

A considerable number of geographers have linked the concept of a region to that of the landscape. Sorre stated explicitly that 'the region is the areal extent of a geographical landscape' (1957), and Gribaudi used almost identical words in demonstrating how much geography has gained and still has to gain 'from an understanding and scientific explanation of integrated terrestrial reality – the region and the landscape' (1965). The straightforward concept of the landscape is of obvious relevance and Despois among others, while broadly accepting the above views, felt it necessary to elaborate this: 'the concept of the geographical region is far more comprehensive than that of the natural region, since it refers to more or less densely settled areas and to diverse and changing economic and demographic structures' (1965).

Numerous Anglo-Saxon scholars should be included in this school of geographers who are concerned with the parallelism between visual and structural characteristics. As early as 1952, Preston James had emphasised the importance of the landscape and of its evolution, and later developed the concept of 'kinetic regions' in which movements and flows were the dominant elements. In his reflections on the idea of the region, Gilbert (1960) reviewed works by English and German scholars but then went on to consider nineteenth-century novelists – the Brontë sisters, George Eliot, Thomas Hardy and Arnold Bennett – who with subtlety and insight had described the English countryside, the setting for the escapades of their heroes. He suggested that 'the task of the regional geographer can resemble in some respects that of the regional novelist. The regional geographer strives to integrate the multitude of seemingly disconnected facts about nature and man in the region he is describing'.

The definition of a region proposed on the occasion of the fiftieth anniversary of the Association of American Geographers (AAG) is worth quoting: 'the region is an area of any size that is homogeneous in terms of specific criteria and is distinguished from surrounding regions by a particular grouping of such criteria, forming a complex inscribed on the Earth's surface . . . any segment or portion of the Earth's surface is a region if it is homogeneous in terms of such an areal grouping'. The full text from which this extract is taken introduces two further important concepts. Not all phenomena present in a region are relevant, only those 'required to express or illuminate a particular grouping, areally cohesive. So defined, a region is not an object, either self-determined or nature-given. It is an intellectual concept, an entity . . . created by the selection of certain features . . . and by the disregard of certain others', in order to study groupings of complex phenomena within an areal unit (Whittesley, 1954).

We should examine the implications of these statements. They are a far cry from a feature indelibly inscribed on the Earth's surface, to be defined and interpreted by the geographer. His task is now to select those elements which seem to him to be fundamental or dominant, and to focus his attention on discovering their causes and consequences, and on dismantling the whole complex of interrelationships. This sequence of cause and effect must at some stage relate to features inscribed on the Earth's surface, and it is on these grounds that we can talk of the landscape as 'the proper perspective of the geographer'.

In a mountainous region, the geographer will obviously select the presence of the highland in some form – either in terms of the type of stockrearing practised (which may also be seen in relation to the area colonised by a particular racial group such as the Salers in the Massif Central, or to possibilities for commercial development such as milk production in the hinterland of the Côte d'Azur), or of the extension of peripheral industrial areas, as in the Pennines, where the textile industries of Yorkshire and Lancashire have encroached on the highland from both east and west to such an extent that workshops following the line of the valleys on either side almost meet. Features of the natural environment are significant in both cases: in the first example, however, they are the major elements to be described and analysed, especially in relation to their influence on human activity, whereas in the second they are only a more or less permissive substratum. Similarly, in trying to describe the vast individual regions of the United States, geographers may consider the Appalachians simply as a highland area or 'physiographic region', or as comprising two subregions – the north, which is linked economically to the Industrial Belt and where towns bring life to the valleys, and the south, which is integrated with the southern States and penetrated by their fields of tobacco, maize and, until recently, cotton. Each of these subdivisions is a distinct and valid unit and each reflects a single dominant characteristic. This underlines the futility of academic dogmatism over the meaning or significance of the word region.

Many authors have tried to reduce the inevitable confusion by qualifying the term with various epithets – such as natural regions, which express broad groupings of physical characteristics; human regions which, according to Le Lannou, reflect the pattern of population distribution and also suggest groupings 'of which we must analyse the internal structure and dynamic interrelationships' (1949); historical regions, defined by former political boundaries, within which a distinctive social and economic environment could develop and of which certain features have persisted despite more recent upheavals. There is also a whole series of specialised regions – economic (to which we shall return), agricultural, industrial, urban, the range is infinite. There is even the phrase 'geographical region'. Here the confusion commences all over again since to non-geographers this has come to be almost synonymous with a natural region, while to others it is simply an area where human activity is the dominant characteristic. However, as James quite justifiably claims, the adjective from the word geography must include all issues relevant to the subject as a whole.

In certain countries the conflict has already developed into an open confrontation. The *Grande Encyclopédie Soviétique* reduced the whole of geography to its physical aspects: however, a new journal published by the University of Moscow in 1958 immediately took a stand against this attitude, claiming that 'the most important task of the science of geography at its present stage of development is the integrated study of regions'. Soviet authors, despite heated debate over the most satisfactory framework for geographical analysis, have maintained an interest in both natural features, historical criteria and the complexities of economic activity. This broad perspective, defended by several academicians including Sotchava, would seem perfectly logical: 'subdivisions of the geographical environment are determined by the particular groups of components selected, while evolution of the components themselves is an interdependent process. As conditions change, they no longer evolve at the same speed or develop the same characteristics' (1956). Sotchava suggested a range of terms to avoid confusion or ill-considered use of the word region and, after reviewing numerous Soviet publications, proposed a hierarchical system of spatial subdivisions: at the lowest level are individual localities, groups of localities and districts; the second order includes divisions, provinces and zones; the highest order comprises geographical regions and the major 'belts' of the world. Rodoman, while adopting the same criteria to define regions 'in a taxonomic sense, as units of a certain order within a hierarchical system', also added a more general interpretation which is still considered valid by many geographers – 'any circumscribed territorial unit'.

His view is similar to that of classical geography where a region was interpreted in a very broad sense, sometimes with a series of qualifying adjectives. Brunhes (1956) defined geographical regions as areas of varying size, each of which possessed a number of broadly homogeneous or analogous characteristics: they were therefore primarily natural regions. Historical regions on the other hand, which were determined by processes of human decision-making, usually comprised several of these physical units. Gallois (1908) shared this view and initially distinguished regions based on physical features, forming units which could then be grouped into economic or historical regions. Lefèvre reviewed similar concepts in numerous articles, and her own approach was clearly summarised in one of her last publications (1965).

Under the influence of planners and administrators, however, certain geographers have given the term a more restricted and specialised meaning. This is true of Gottmann, whose definition is clearly related to the politics of regional planning: 'the region is a complex created by man and which man can destroy' (1952). For many other scholars including George, Kayser and Rochefort, the concept of the region also hinges on a capacity for organisation. Regions are simply spatial units with a clear and often polarised structure, dependent on a series of nodal points. George stresses that a region or system of regions is associated with the urban network. Kayser, studying embryonic regionalisation in underdeveloped areas, distinguishes 'undifferentiated space' which lacks clear economic subdivisions, and 'specialised space', an area with a certain type, level and diversity of economic activity, to which he liberally and consistently assigns the term region – for example, to the area of impact of commercial enterprises, such as plantations or industries established round an urban centre with concentrated capital investment, or to areas structured by administrative and political considerations.

It would obviously be wrong to assume that there is no relationship between the various types of regions – the developing areas provide conclusive evidence of such relationships, as also do the processes of historical evolution to which geographers so often refer. One can trace the development of nodal points, the initial stimulus to which has often been external, in the form of a colonial administration or of commercial and social investment. Contact between two very different levels of economic and technical efficiency is reflected in diversified and accelerated development, diffusing like an oil stain over a more or less extensive area. But there the analogy ends. The impact of plantations, or of an area irrigated from a large reservoir, may give rise to a uniform region neatly inserted into the traditional surroundings, whereas the sphere of influence of a town or an administrative structure will show an unequal and variable diffusion pattern, including a distance-decay function – it will exhibit 'polarisation' to use the current terminology.

On the other hand, if within the same general area there is a clearly-defined plain, distinguishable from the neighbouring plateaux by its soils and climate and with diversified economic activities, does it not also have a claim to the label 'region'? I am thinking of the Reconcavo de Bahia, a pronounced sedimentary rift valley lying between two crystalline blocks, where a succession of landscapes from the marine littoral to the interior

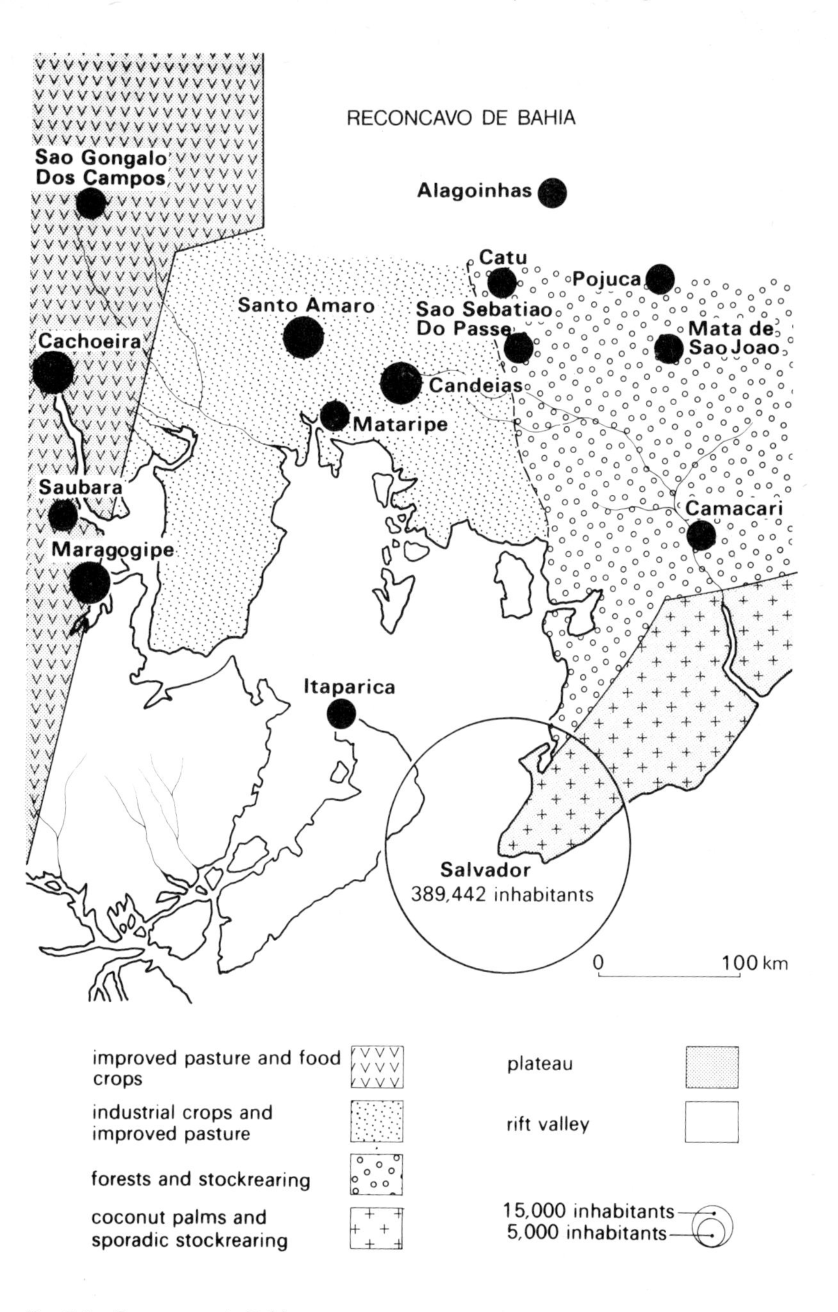

Fig. 3.5 Reconcavo de Bahia
A sedimentary rift valley, distinctive in its rock type, agricultural economy, and small towns expanding rapidly as a result of the discovery of oil

reflects the nature of the sedimentary outcrops and variations in humidity – from the dense humid forests to the grazing lands and sugar cane plantations. The influence of the large city of Salvador, situated towards the southeast margin of the valley, was scarcely discernible twenty years ago but the Reconcavo formed a distinct unit in terms of its population distribution, its dense network of market towns and its agricultural system – this was clearly visible from the higher crystalline area to the west. It is still not so much the influence of Salvador as the discovery of oil reserves underlying the valley that has been crucial to the economic transformation of the region. Although the offices of the Compagnie Nationale are located in the city, the refinery and export terminals, which have modified the environment and brought both a revolution in industrial technology and an influx of population, are situated where the extreme southern margin of the plain opens on to the bay of Todos os Santos. Oil wells now line the valley, while company employees have transformed the small towns with their new activities and high salaries. The individuality of the rift valley is obvious in its external features, and in both the agricultural and industrial sectors these are related to a structural accident.

This one example from many is evidence that even in a developing country the division of space is not a simple issue. It is not based solely on a limited number of structural elements, whether economic or political, nor is the environment in general undifferentiated: it has its own set of rules for division reflecting its own natural features, the impact of its past history and ethnic variation. Pélissier and Sautter found striking examples of 'traditional regions' in West Africa, and they can be identified without ambiguity over much of the remainder of the continent. They are characterised by a relatively high density of population, 'associated either with a clearly-defined natural environment or with a distinctive ethnic or geographical personality (as in the case of the Sérères, for example), or with both of these features (the Bamiléké plateau . . .) relationships which are everywhere the outcome of a slow process of adjustment over both time and space' (Sautter, 1967).

Certain geographers regrettably dismiss the details of this complex substratum and examine only economic and administrative structures – primarily urban influence – thereby immediately oversimplifying the framework for spatial analysis: they neglect considerations of greater depth, which are obviously also the least apparent and therefore take longer to appreciate.

It is perfectly legitimate to identify different categories of regional division. It is also essential, particularly in relation to planning, to specify structural elements that are relevant to modern societies, although it would seem unjustifiable to reserve the term region exclusively for spatial units so defined. A region would then acquire a precise but limited meaning and one which, according to the most authoritative dictionaries of the French language, is also erroneous.

Regions and regionalisation

We have still not exhausted the definitions of a region. Claval, working through a mass of specialised documents, devoted a large volume to the question of regionalisation and suggested that regions should be considered as 'spatial structures which are smaller in area that the State, which possess a certain individuality, and which are considered as an entity either by the people who live there or by outside observers' (1968). This definition recurs in numerous official documents although rather less explicitly. 'A region – the highest level below that of the State at which the various forces influencing social and economic life interact' (Motte, 1960). This leads us into the world of administrators and planners. 'In the eyes of regional planners, the prime objective of regionalisation is to provide a framework rather than to define its contents' (Quermonne, 1964).

No-one would deny the validity of this type of region. It is a limited but long accepted usage of the Latin word *regio*, which originally referred to a straight line delimiting the regions of the heavens and drawn by the Augurs in their search for omens. It was rapidly adopted by administrators, and used to define the territorial units of the Roman Empire established by Augustus to facilitate census-taking and tax collection. It was also used within Rome itself as the equivalent of a quarter or district. And on an international scale, military, judicial and fiscal regions have long existed without seeking permission or approval from geographers!

Discussion was re-opened in France when, for reasons of planning efficiency, the government arbitrarily divided the country into twenty-one economic regions. This policy merits further elaboration since it highlights our problem. The regions comprise a variable number of départements and have no common basis of delimitation. The Midi-Pyrénées, for example, includes twelve départements and inherited a fragment of the Massif Central, the plains and hills of Centre Aquitaine and the Hautes Pyrénées: the Loire valley and its margins, on the other hand, are divided between the Centre and the Pays de la Loire. The new regions lack the historical basis of the German Länder and the cardinal framework of English economic regions, areas which in both cases are considerably larger. Without escaping from them completely they attempt to break away from morphological units, which as a result have often been arbitrarily subdivided: Languedoc–Roussillon and Provence–Côte d'Azur share the Mediterranean littoral, Auvergne and Limousin the centre of the Massif Central, while the Paris Basin has lost the whole of its outer margin. These economic regions bear no relationship whatsoever to any traditional divisions of France, although in some cases they acknowledge historical boundaries (Burgundy, Normandy) and in others modern industrial agglomerations (Nord). They have been criticised by several geographers and Labasse (1960, 1966) in particular has been highly outspoken.

Leaving aside the pointless quarrel over tradition and returning to the dual properties of geographical space – what is our verdict on this attempt at regionalisation?

However heterogeneous in geographical terms, these divisions are useful from an economic and administrative point of view, since they effectively delimit an area of variable and sometimes complementary characteristics around a town or centre of economic activity. In some cases, more or less homogeneous natural regions have been subjected to complementary economic forces, operating from within and from outside their boundaries, precisely because their inherent potential was uniformly distributed. The margins of the Limousin, for example, were developed together with the peripheral plains, Roanne and its textile industry were and still are closely linked to Lyon, the market towns of distinctive agricultural areas such as the Beauce or Brie are peripherally located, and are often in close contact with very different rural economies. The two margins of the Vosges relate economically to their adjacent plains, as does the eastern edge of the Massif Central to the Rhône–Saône valley. Demographic and economic breaks mark the outer boundaries while the rural population and towns are concentrated along the valleys, especially at prominent sites such as a confluence or an estuary.

To delimit economic regions on the basis of a growth pole, and of a surrounding area of economic activity proportional in size to the propulsive strength of the pole, is a valuable technique in regionalisation for planning purposes. Where regional policies are flexible, this strategy might be judged a sound one – provided that the growth pole has a favourable central location, a broadly convergent communications network, a satisfactory level of amenities, and administrative and financial resources commensurate with the role assigned to it. In the French situation, however, the small size of the regions, the lack of a focal point, and the inadequate resources available to the central administration are serious weaknesses, and to have drawn up a policy of decentralisation and regionalisation on such a basis was both hypocritical and inefficient balkanisation.

Demythologising the word region

We should conclude our discussion by affirming the necessity to demythologise the word region. We should either decide to use it in the vague sense of an area or space, consistently adding an epithet to specify a natural, historical, economic or planning region, or use the word on its own to refer to an administrative region, a politico-economic unit just below the national level in the hierarchy. However, since there is little hope that military, judicial and many other regions will simply disappear at our command, we should dismiss it as trivial and abandon a pointless battle of words. Instead, we should consider how the geographer can conceive, describe and formulate his spatial divisions in a sufficiently realistic way to achieve his overriding objectives.

4

The division of space

We must first clarify the meaning of the expression 'division of space'. A somewhat rudimentary division may provide a convenient frame of reference for studying particular groups of elements. A geomorphologist, for example, examines contrasts in relief type: he analyses genetic relationships with geological structures and past or present climates and also describes the various morphological regions, subdividing them as he thinks best. A population geographer will probably consider distribution as the most significant factor: he will contrast nodes and axes, or areas of high and low density, and discuss the implications of the distribution pattern. These are certainly practical and useful divisions, but do they really constitute the 'geographical regions' which should be the culmination of our research and an original contribution to the understanding of our planet?

A demographer also calculates population densities and presents their distribution in cartographic form, but a difference of approach is already apparent. The demographer is content to make use of administrative divisions and his maps are almost exclusively of density, that is, of an even distribution of population with respect to a particular area. Geographers, on the other hand, frequently employ a technique that is just as appropriate and more realistic – representation by a point pattern. They alter the spatial framework and substitute variation in the natural environment for the artificial unity of an administrative area. The fertile valley lying between two arid plateaux is then reflected in the population pattern, while the marshy lowlands appear deserted in comparison with the line of hills or the well-drained plateaux along their margins. Any framework which ignores certain spatial factors is therefore inadequate for a valid geographical analysis.

If only for practical reasons, however, total space cannot be comprehended without a framework and the principles of subdivision must be determined on a geographical basis, according to general combinations of elements rather than their individual characteristics. An area in which only the geomorphology is analysed is a specialised geomorphological unit, not a geographical region. If an area is uninhabited and devoid of vegetation, as in a desert, its morphology simply reflects the balance between opposing forces – a given geological structure and climatic conditions – but in the majority of cases such 'naked' relief does not exist. Similarly, the area of influence of an urban centre is not a geographical division if it is defined solely by linkages which ignore the diversity of the static substratum, that is, variation in both natural and human environments.

Analysis of individual elements and of their spatial expression is nevertheless essential to the identification of geographical regions. The only way to determine significant combinations of elements is first to delimit areas on the Earth's surface by a series of relatively straightforward static criteria, and then to assess by various means the extent to which they coincide with areas of complex dynamic relationships. A cohesive grouping of elements relates both to the area itself and to its limits: an area should therefore be defined only on the basis of characteristics common to all its elements, clearly distinguishing them from those of neighbouring spaces. This introduces two further concepts – that of a region which comprises identical elements and defines a homogeneous space, and of one based on complex and specialised relationships between complementary elements. Homogeneity always implies equality and therefore a juxtaposition of elements too alike for exchanges to develop between them. Complementarity, on the other hand, immediately suggests the idea of association. This may take many forms: a valley and the surrounding plateaux are often complementary in both physical and economic terms, as were areas of stockrearing (La Perche) or of viticulture (Val du Loir) with those of cereal cultivation (the Beauce) in the days of a pre-industrial rural economy. Also complementary are a port and the hinterland producing export commodities, a rural district and the town providing it with administrative, commercial and banking services.

The generation of a whole spectrum of complementary flows and relationships often finds immediate expression in the landscape, although its description and precise evaluation are far from easy. Homogeneity, on the other hand, is relatively simple to identify at the macro-scale, but so subtle are the local variations that it seems to defy detailed analysis.

Brunet (1969) attempted to identify for rural areas the smallest practical units containing identical elements – his so-called 'rural districts'. If we examine all the characteristics of the physical environment, economy, population, linkages and production, areas such as the Aquitaine region of the Midi, which at first sight would seem to contain a number of extensive uniform areas, are reduced to a micro-mosaic in which each unit is totally homogeneous but comprises only two or three communes. Two obvious points arise from this and other similar examples – the likelihood of homogeneity increases as the size of the area is reduced; the homogeneous area will be more extensive if very general criteria are adopted. If the sole criterion is a plateau, all plateaux will fall into the same category; if it is plateaux coinciding with a resistant structural surface, their number is already reduced; if we then add the presence of surface clay deposits, fewer still will be included, and so on. In dealing with the complexity of human factors it is easy to understand how many geographers, though perhaps with an excess of zeal, have ultimately come to proclaim the concept of uniqueness.

Spatial divisions can therefore be determined only after research into the underlying elements – I am tempted to write 'into the units of measurement'. The grouping of elements will require greater skill if the elements themselves possess a highly complex vertical structure, in which case the area is likely to be small. Uninhabited environments are less liable to fragmentation than those where the presence of man has superimposed increasingly sophisticated and complex settlement patterns (Kolotievskij, 1967). A basic guideline for research into factors governing the fragmentation of space is to examine the role of natural elements before that of human societies, in other words, first to establish general rules for the definition of boundaries and of homogeneous areas. Superimposed on the resultant mosaic is a more or less dense lattice of dynamic relationships, indicating further divisions based on linkages between various sections of the mosaic, that is, on the organisation of space.

Physiographic division

In all the examples discussed so far, there have been numerous implicit or explicit references to the physical environment. When an observer surveys an area from a high vantage point, whether or not he is a geographer, he is first struck by the contrast between lowlands and mountains, between the valleys and the summits or, if he were to fly over the Nile delta or a desert, between the green of the irrigated areas and the grey-white aridity of their borders.

While the role of the physical environment cannot be neglected in any approach that calls itself geographical, it is of paramount importance in the division of space. 'I believe it is within nature itself that we should seek the principles for geographical division' (Gallois, 1908) – a significant but perhaps exaggerated statement which we might qualify with the more restrained words of Gourou: 'physical factors exercise an influence only as a function of the civilisation which interprets them'. Man can submit or adapt to physical conditions, he can overcome or transform them to some extent but can never ignore them: they present him with an inescapable framework, ranging from straightforward possibilities to dramatic constraints. But the natural environment is itself complex: it comprises intricate relationships between diverse elements, in such a multiplicity of combinations that we should need to describe hundreds of type situations in order to appreciate its influence in every case. Temperate lowlands composed of permeable or impermeable soils, in proximity to the sea or to a mountain chain or crossed by a river, possess neither an identical physiognomy nor potential for development. We are therefore forced to look for type-characteristics and to indicate their hypothetical role, rather than become submerged in the infinite variety of individual cases.

The primary elements, combinations of which determine the nature of the physical environment, are climate, relief and soil type: the natural vegetation broadly reflects their interaction – although not all are of equal significance in the division of space (Sotchava, 1956; Lehmann, 1967). To confine ourselves to relatively straightforward considerations however – climate determines general zonal patterns while relief is the basic agent of diversification, directly or indirectly the true creator of an environment for living.

Both slope and altitude literally dissect certain areas of the Earth's surface into hollows, basins, troughs and furrows, all of which are favourable locations in terms of their aspect, their pattern of convergent slopes and parallel valleys, their sheltered climate, the concentration of water, mineral and energy resources, and of a labour force originating in the nearby highlands for which they are the natural outlet. The contrast is even more clear-cut between areas of high relief and the lowland elements which dissect or border them. An outstanding example is Japan, where the mountainous backbone has defined the major regions of the country, and where these have subsequently been further differentiated through their use and development by man.

A hypsometric map of the world illustrates the different orders of relief. The great land masses of North and South America, Africa, Australia and central Eurasia, and other extensive highland or lowland areas where successive waves of human settlement have ignored man-made boundaries across vast uniform landscapes, contrast with the deeply dissected island arcs and peninsulas of the Asiatic fringe, and the north-west extremity of ancient Europe with its indented oceanic and lacustrine shorelines. Political frontiers often closely follow a line of mountains but oscillate across lowland areas. The upheavals in Poland following contact with the Slav and Germanic empires of Europe, the prolonged struggle over the division of the Chaco between Bolivia and Paraguay, the geometrical carving up of the centre of new continents which lack a pronounced mountain barrier along lines of latitude and longitude are all evidence of the function of relief as a framework for human settlement.

This is not of course the same as an automatic determinism which would turn all highlands into demarcation lines and all mountains into areas of repulsion in relation to neighbouring plains, except under locally favourable climatic conditions. Frequently does not imply always or invariably. A mountain chain may constitute a barrier or a unifying feature. Since the Second World War, for example, the political frontier between France and Italy has coincided with the summit line of the Franco–Italian Alps, whereas before 1860 the whole area formed the northern nucleus of Piedmont–Sardinia: this kingdom was guardian of the passes linking both flanks of the mountains. A relatively hostile highland area may also support a higher density of population than the infinitely more attractive neighbouring plains, as on the rugged and beautiful Kabylie in Algeria and the lateritic Fouta Djallon of Guinea: here man has taken refuge against man, as on the high Andean plateaux he found protection against the malevolent lowland climate – hot, humid and a dangerous breeding ground for malaria.

To these multiple and subtle combinations of slopes, aspects and altitudes should be added the influence of the underlying rock, surface deposits and soils, all of which also relate to climatic fluctuations. The vegetation type reflects the interaction of these factors and is therefore a useful indicator in the division of space. Natural vegetation might be considered as the reactive of morpho-climatic conditions – but in how many regions of the world, other than in climatic zones inhospitable to man, does it exist in this form? What is its role in regions where, as a result of vigorous human activity over a long period of time, the degraded natural vegetation forms only residual elements? It constitutes small islands in the midst of cultivated lowlands, a refuge for recreation and hunting, and is generally an insignificant fragment in terms of our overall perspective. Certain forest areas however, such as the Landes and the high Jura, may be so extensive that they either themselves constitute a separate region or serve to reinforce the limits of an existing natural division.

Westward colonisation of the United States was deterred for many decades by a combination of obstacles, in particular by the outer ranges of the Appalachian fold mountains and the dense forests which provided cover for hostile Indian tribes. The centre of the continent was first penetrated via the waterways of the St Lawrence, the Great Lakes and the Mississippi. This illustrates the relative importance of various environmental factors in the organisation of space. An accessible coastline had attracted the settlements of the first colonists, and their survival was due both to the maintenance of links with the home country and to the exploitation of local natural resources. The barriers to further colonisation were the forested highlands and the enormous distances to be covered, and the more accessible grasslands were therefore opened up via the waterways. Even at the present time, these natural features form conspicuous dividing lines between the major regions of North America.

This example would seem to support the contention of several authors that water is one of the most significant factors governing spatial organisation (Labasse, 1966). It is obviously a basic element: on the lack or availability of an adequate water supply depends the absence or concentration of men. Variation in the nature of the catchment area, the flow, water table and springs leads to marked differences in the possibilities or problems challenging man's technological resources. Water, whether falling from the skies, present on the surface of the Earth or in subterranean reserves, offers inestimable potential wealth which will increase still further when desalination techniques have been perfected.

The role of the various physical elements in the division of space is therefore not as straightforward as a purely deterministic viewpoint might lead us to believe. Three important concepts emerge from the complexity of patterns and constraints – the nature of the environment, its susceptibility to human activity and the scale of analysis.

At the zonal scale climate is the dominant factor; as in the major regions of Africa or Brazil. Similarly, the influence of the temperate oceanic climate in northwest Europe is felt across the plains as well as over the hills and mountains. The same is true in Mediterranean countries, where a landscape comprising barren arid regions, scrubland and evergreen trees is found from Provence to Lebanon and from the Peloponnese to Cap Bon, despite the physical fragmentation of the lowlands and the ruggedness of the surrounding mountain arcs. However, the imposing mass of an unbroken line of mountains often suggests that relief is the dominant factor. In the Andes or the high plateaux of the western United States, for example, one instinctively considers relief before the numerous climatic variations which reflect their altitude and latitude.

There would therefore seem to be a reversal of the role of climate and relief at a more detailed scale of analysis. In Scotland, for example – one of the more northerly temperate oceanic regions – the environment for human life in the Highlands and Lowlands is very different: even within the Highlands the western slopes are steeper and more exposed to maritime influence, while the gentler eastern slopes are sheltered from prevailing winds. The highlands of the western states of Washington and Montana are dissected into a series of crests, longitudinal valleys and intermontane plateaux: the western slopes and overhanging summits are well-watered and wooded, while the eastern slopes are far more arid. Man has colonised vast areas and developed a wide range of farming systems in response to variations in temperature and humidity.

At a still more detailed scale potential subdivisions are almost infinite, of which the geometrical pattern of the Franco–Italian Alps affords an excellent example. They have a longer slope towards France and a shorter one towards Italy: the steep southern margin of the Massif du Pelvoux and the valley of the Drôme subdivide the French slope into two basic units – the northern and the southern Alps. The northern Alps are characterised by a fourfold longitudinal division – the Pre-Alps, the Sub-Alpine Furrow, the Intra-Alpine Zone and the High Alps. The Intra-Alpine Zone is dissected into a series of troughs by numerous transverse valleys, including the Maurienne. One could analyse the morphology of this valley – the succession of villages, the system of agriculture, the alpine pastures on the *adret*, the meadows and forests on the *ubac*, below the rock walls and the eternal snows.

At what scale, however, is a significant geographical subdivision to be found? To determine this we must reverse the above exercise. The vertical succession of land use refers to the Maurienne, but the Tarentaise and Romanche exhibit a similar pattern: the valley floor and slopes, together with the surrounding summits, form a single morphological unit. Despite minor variations, the other valleys of the Intra-Alpine Zone also fall within this broadly homogeneous area, itself linked geologically, climatically and hydrologically to the ranges of the High Alps. The latter have two slopes – one is the Intra-Alpine Zone itself and the other directly borders the southern Alps – while beyond are the lowlands of the Po to the east and the Rhône valley to the west, in this case a complementary rather than a homogeneous association of physical features. The subdivisions of the northern Alps contrast in almost every detail with those of the southern Alps, and the regularity of the four parallel zones also disappears beyond the Swiss frontier and the trough of Lake Geneva. To my mind, the northern Alps constitute a truly significant natural geographical region, comprising several sub-regions which are themselves complex and interdependent: even if entirely uninhabited it would still confront surrounding regions with a striking individuality. It is only fair to add, however, that a map of population distribution would emphasise the fragmentation of the region rather than its cohesion, by contrasting the populated valleys and uninhabited summits – there is little evidence of an essential complementarity in terms of the human environment.

An analogous situation is that of the Saône lowlands. Although the Dombes, Bresse and the Chalonnais do not resemble each other in every detail, this succession of lowlands is clearly delimited by the relatively uniform surrounding highlands. The eastern margin of the Rhône valley above Lyon marks a change in relief type and the curve of the limestone

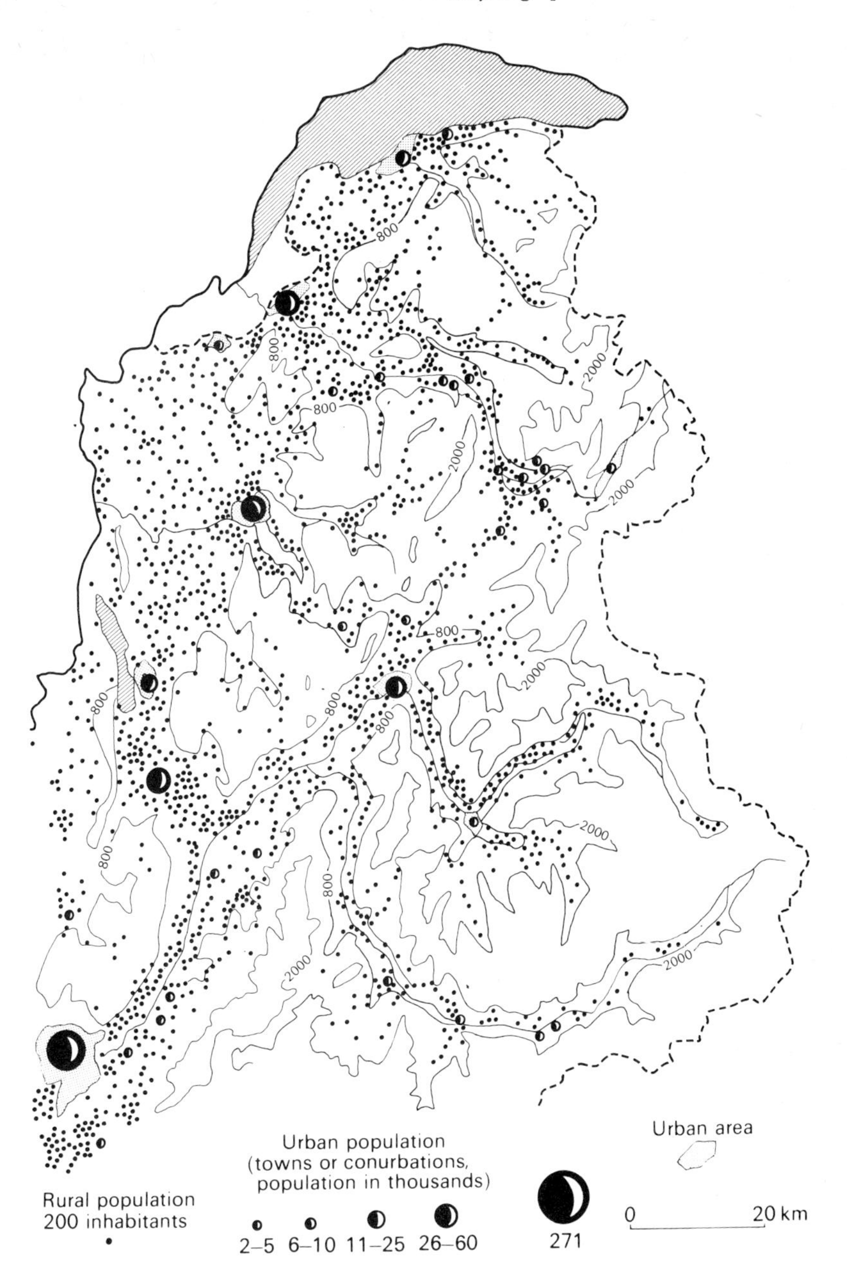

Fig. 4.1 The distribution of population in the Alpes du Nord
This map illustrates both a technique of representing population distribution used by geographers and also its application in defining the various units within a region

plateau of Burgundy and Lorraine almost closes the circle to the north – beyond which the valley of the meandering Saône forms a natural region of a different type, but with limits also marked by significant changes in relief. There is a greater difference, for example, between the edge of the Jura and the Dombes than between the Dombes and Bresse.

These are obviously simple situations which have deliberately been chosen to illustrate a particular point of view. Physical boundaries are often indeterminate or in dispute, but we should still attempt to identify and above all to justify them. This again evokes the classical concept of a 'natural region', but I would prefer the phrase a 'geographically significant natural environment'.

Such space may well not exhibit absolute unity in the strictest sense, but possesses its own coherence and imposes a series of constraints on human activity. It therefore has fundamental geographical implications, since all human societies colonising a particular region will have to face the same range of problems.

This coherence can obviously occur at different levels. Although we always refer to the Paris Basin, for example, farmers of the Beauce do not contend with the same basic conditions as those of the Champagne or the Vexin of Normandy, farmers on the margins of the Ile-de-France as those of the plateaux of Brie or Champagne, of the valley of the Somme as those of Picardy. While this may be true, and local variations have been acknowledged in the identification of 'pays', far more striking contrasts are apparent between inhabitants of Brie and of the Vosges, between a farmer of the Beauce and a stock breeder on the bocage of Normandy or on the Morvan. We must also consider natural focal points. Convergence of the drainage network within the Paris Basin encouraged an associated convergence of lines of communication and specialised linkages: there is no need to point out the location of Paris to demonstrate this. Similarly, if we delete the city of Lyon from a map, the convergence of natural routeways is still an outstanding feature. The physical environment is not merely a spatial relationship between more or less interdependent factors: it is an interrelated whole, a self-contained and distinctive grouping of elements.

We return therefore to the concept of the relativity of environments – both relativity of dimensions, since the 'cells' of Europe for example have little in common with those of Africa or America, and relativity of environmental conditions. It would be absurd to deny that the environment does impose pre-determined constraints. Certain environments are inflexible – despite man's technological progress the Himalayas will always constitute a major barrier, and the Pyrenees a minor barrier of the same type: the Arabian and Saharan deserts or Greenland will never be more than vast open spaces settled only sporadically and at great cost. Other environments are adaptable, since they either lack extremes of physical conditions or their size and location permit a radical transformation – such as the Negev, the coastal lands of the Netherlands formerly below sea level, and the Landes of southwest France. There are also quasi-neutral environments where it is possible to realise a wide range of projects, and favourable environments endowed with a mild climate, fertile soils and an ideal combination of all their elements. The concept of pre-determined conditions is a passive one, however, and does not take into account the degree of adaptation by man to his environment. For example, 'the Mediterranean coast of Africa, although the worst endowed with natural harbours, has fully participated in the general economic life of the Mediterranean region . . . the estuary of the river Gambia, although navigable up to a distance of 200 km from the sea, has not given rise to any notable maritime activity . . . the hostility of the African shores is perhaps an exaggeration arising from the absence of trading activities among its coastal populations' (Gourou, 1970).

It would be erroneous and totally unrealistic for a geographer to ignore this basic human factor in any research into the division of space, its use and organisation by human societies.

The influence of human activity

The possibilities for human activity within this varied and variable environmental framework are almost infinite. Several authors have devoted long volumes to this one theme – Isard (1956, 1960), Labasse (1966), Claval (1968), Berry and Marble (1968) – to mention only a few relatively recent methodological works, each of which contains a substantial bibliography. However, I propose to adopt a very simple framework within which to identify common themes and to outline a regional typology.

Certain authors claim that evidence of human activity is a prerequisite for use of the term 'natural region'. Sautter (1966) discussed this in detail in his thesis – 'we should hesitate to apply the term "natural region" to an area which is simply the crude expression of the physical environment' – and suggested the following definition: 'an integrated area, delimited on the basis of human activity in an environment with clearly-defined physical features . . . it is only in the sense of nature as interpreted by man that we can refer to a "natural region" '. Le Lannou (1949) had expressed a similar view in claiming that the Paris Basin was a natural region only because human activity had served to emphasise its distinctive physical characteristics. This view might be considered too anthropocentric but it reflects an undeniable truth that the interaction between man and his environment is often so close that it is impossible to separate the consequences of physical and of human transformation. In my opinion, however, the adjective 'natural' should refer exclusively to a physical environment unaltered by man. Once man intervenes, however superficial or limited his activity, we must use the term 'geographical' in order to be consistent with our definition of the discipline itself. A physical space may also be 'geographical' in environments where human activity has been only sporadic or intermittent.

The observations by Sautter and Le Lannou are interesting in that they again revive controversies – which we had dismissed as pointless – over the word region. They deserve further consideration, however, since they develop the concept of 'a geographically significant physical environment' modified by human activity – although we should examine only those forms of human intervention which result in a distinctive spatial coherence. In other words, we should focus our attention not on micro-features or random occurrences but on those which exhibit a regular pattern and which enable us to define regions of varying size and systems of relationships.

For example, a recluse who clears a corner of the forest for his log cabin and garden certainly creates a micro-division of space, but this should be included in a typology of forest clearance or of the activities of pioneers: it is not of interest to a specialist geographer in his search for significant spatial units. On the other hand, if a group of settlers were to penetrate an area via the river valleys and make a series of clearings along tracks at right angles to them, as in the Canadian 'lines' studied by Deffontaines (1953), the network of paths and huts and other appendages of colonial settlement would certainly constitute a geographically significant environment, and could be considered as an element in the division of space. There is a physical framework (a forest dissected by rivers), a human community pursuing a unique way of life, and a distinctive mode of spatial conquest with expression in the landscape. The numerous interrelationships clearly define spatial units relevant to geographers.

Suppose that a group of migrants has colonised a vast area – as in the above example. We can refer to this area of settlement as a zone and map its total extent. Several rivers dissect the zone, however, each with its own network of tributaries: the forest constitutes a barrier on the landward side, and the isolation of the 'lines' is broken only by communication along the waterways.* We would normally expect the 'lines' within network A to possess stronger internal than external linkages and that their inhabitants would be in contact with each other, meeting occasionally in the most important settlement located at a confluence . . . the same would hold true within networks B, C, . . . The physical environment therefore intervenes twice – through its broad zonal characteristics and through the river basins which it demarcates. However, once the inhabitants of A have overcome physical constraints and established an overland route to a point within B, zonal subdivisions should be based on new relationships between the two groups of settlers: this will result in an area of interaction (A + B) and one of limited relationships within network (C).

Conversely, imagine that network D is crossed by a political frontier: the headwaters are in Canada and downstream waters in the United States. There is therefore a political and administrative division of a zone of settlements with broadly similar physical characteristics. Since legislative policies are also likely to vary between the two countries, other divisive features will become apparent: different languages may be used, trade will not be entirely without restriction, and separate supply centres will serve the Canadian and American districts. This area, selected primarily to illustrate a broad zonal framework, could therefore be subdivided in other ways. The 'lines' can be considered as the elements in the analysis and the zone as a static unit, while its administrative subdivision and areas of interaction constitute functional units.

Continuing this theoretical approach, we can distinguish three forms of human activity: the immediately visible signs of land clearance, the possibilities for establishing areas of interaction, and the formation of

---

* These simplifying assumptions render the example extremely hypothetical, however.

distinct spatial units. In reality, however, these categories are not so simple, independent and rigid, as will quickly become apparent.

A visual division

A first purely visual inspection of an area would suggest a fundamental contrast between predominantly rural districts and those with a high degree of urbanisation. In rural areas, the most striking feature is the cultivation of the land, the planting of a crop cover or of semi-natural vegetation (improved grassland) for the grazing of stock. Two types of boundary should therefore be easy to delimit – that between cultivated and uncultivated or abandoned land, and the periphery of the urban area. In reality, both are open to criticism from our point of view. Cultivated land may 'disappear' without relating to the outer boundaries of a particular rural community: under early economic systems, for example, the productive land of villages in the Alpine valleys included high pastures beyond the fields of the lower slopes. Similarly, it is often difficult in African countries to distinguish between unfired and regenerating forest when, several years after initial burning of the vegetation, the soil has recovered its fertility. These two examples are simply indicative of the innumerable practical difficulties which may arise.

We must also consider the selection of criteria for delimiting rural areas: should we examine the proportion of cultivated land, the relative percentages of arable and grassland, the density of the population employed in agriculture, the type of rural settlement, or rather the judicious combination of a range of criteria, carefully selected to ensure the most accurate description of the environment? I shall return to this point. For the moment, let us broadly accept the suitability of rural environments *per se* – given certain limitations – as a possible basis for subdivision of the spatial mosaic.

In a purely visual sense, similar problems arise with urban environments. We shall consider only spatial elements, rather than the finer points of the nature of urbanisation and various criteria adopted to define an urban centre. Since we are dealing not with point patterns but with extensive surface areas, two or three examples will suffice: an 'urban nebula', as on the coalfield of Nord–Pas-de-Calais and in areas where coal or metalliferous deposits give rise to islands of activity quite distinct from the surrounding rural landscape; conurbations, such as the monsters of the Ruhr, Lancashire and Yorkshire; a megalopolis, or rather a polymegalopolis, the most famous of which are the central eastern seaboard of the United States, including Boston, New York, Philadelphia, Baltimore and Washington, and the Tokaïdo megalopolis of Japan, including Tokyo, Nagoya, Osaka and Kyoto. There exists a whole range of urban spaces large enough to form clearly-defined areas, some compact and others hydra-like, even within a purely static framework excluding peripheral relationships and influences.

Simple observation therefore permits a crude division of the greater part of the Earth's surface into uninhabited or sparsely-settled areas, relatively homogeneous agricultural areas and well-defined urban zones. This is totally inadequate, however, since it ignores an important section of geography relating to the dynamics of space – from both a static and a functional point of view: vast areas of no man's land, intermediate spaces which do not fit the rural classification outlined above, the omission of a whole range of flows and exchanges.

Administrative divisions and the legacy of the past

A second perspective concerns man's own division of space, primarily along administrative boundaries – whether they delimit regions of varying extent, authority and cohesion within a country, or define the international frontiers of a State. These divisions often relate to physical features such as a river valley or summit line, but in some cases are simply arbitrary or geometrical units. Particularly in countries with a long history of settlement, they reflect the traditions and heritage of centuries: historical influences are highly significant even in the geographical division of space.

The legacy of the past has always been a general concern of French geographers: it is evident in such diverse forms over such vast areas of the globe that it might almost be considered the dominant influence. Research into the origin of urban and rural settlements, studies of political and administrative boundaries at any scale, or of rural land tenure, almost invariably lead to an event in history. But we must not lose sight of the fact that history itself is simply a sequence of relationships, reflecting human activity through time (the nature of the population and its ability to adapt to or transform the natural environment, its capacity for organisation and legislation). Relicts of the past therefore represent a heritage of constraints and possibilities: the geographer must determine why a particular phenomenon appeared, and how and why it came to be preserved. This preservation takes a variety of forms and is relevant at numerous levels of regionalisation.

Firstly, it provides a broad perspective, a canvas on which remnants of past civilisations present a vast panorama. The frontiers of the Roman world, despite its eventual collapse, are still clearly defined along the line of contact with the Slav and Germanic empires: there are also significant differences between Anglo-Saxon and Latin-Indian America. Secondly, the survival of local features may facilitate the delimitation of homogeneous areas (a village farming system), the analysis of complex structures (evolution of a town plan over the centuries) and an understanding of exchange flows and human migrations. In many areas, administrative boundaries acquired such strength in particular periods of history that they became an effective framework for decision-making and for economic development. Many of the regional units thus defined have persisted for centuries. Administrative structures are clearly influential in directing and standardising man's activities. Successive divisions into provinces, départements and economic regions are basically analogous, but with the fundamental distinction that environments formerly operated virtually as closed systems, the diffusion of knowledge and technology was a slow process, and the uniqueness of structures less difficult to preserve.

Claval (1968) has discussed the relationship between historical regions and economically homogeneous zones at the beginning of the modern industrial era in western Europe, and has shown that 'the majority of historical regions attained their present dimensions in the relatively distant past, from the thirteenth to the sixteenth century'. There are many traces of these former regions, but not always in the form of simple and direct linkages. Historical regions were related primarily to the potential for regional economic development (that is, to sources of capital, labour and raw materials), based on primitive forms of industrialisation in the regional capital for a purely local market, and on the dynamism of an individual household of a group of enterprising families. This development process warrants closer examination since in many cases it has a bearing on contemporary structures. Once in existence, a region then 'attracted communities of varying race and religion . . . these communities became interdependent . . . and the region came to be inhabited by a society with similar attitudes' (Claval, 1968). A coherent administrative structure may therefore give rise to a common way of life.

The persistence of this regional cohesion is evident in numerous ways. Many French people feel themselves primarily Bretons or Auvergnats, although for centuries the boundaries of these provinces have ceased to have any significant role: regional studies in France are still generally based on these divisions, one of the best examples of which is undoubtedly Lorraine. In Germany, the present Länder have resolutely preserved boundaries pre-dating negotiations towards the Customs Union. In Latin America the larger cities, State capitals and political frontiers, and therefore the major spatial divisions and their internal organisation, reflect the rudimentary administrative system of the Spanish colonial period, which was itself partially superimposed on centres of Indian settlement.

A common sense of belonging, survival of the boundaries of both single fields and vast provinces, towns periodically rebuilt on the same foundations, and numerous other features all reflect the strength of the forces of continuity. The mere existence of an administrative boundary can set in motion a whole series of processes with long-term consequences. The laws of 1836 and 1865, concerning parish roads and branch railway lines respectively, gave certain départements the opportunity of constructing a convergent road or railway network which reinforced the centralising influence of the main town. The star-shaped network of the Sarthe around Le Mans is a perfect example of this situation. More recently, bus companies have also tended to follow the dictates of administration, through the location of their central garage in the main town and the provision of services only as far as the limits of the département. These are visible signs of the polarisation which has gradually come about in response to official demarcation lines.

Such divisions are not fundamental to geographical study, however, and are often based on a single criterion: they are frequently superimposed on other more significant geographical units, more closely related to the landscape or the economy. It is not because of the boundary separating Essone and Seine-et-Marne that the commercial activities of Corbeil or Melun have had little impact outside their immediate hinterland. The Paris conurbation is a living and coherent reality, although divided among eight départements, and it would be wrong to assume that its influence is only local or superficial.

The delimitation of an administrative boundary is one of the means by which developed societies have sought to impose regional divisions. New states were therefore created as colonisation and settlement of North America advanced: the Revolution which overthrew the national political system in France also led to changes in regional boundaries. To take a small-scale practical example by way of contrast: a wealthy département has a road network superior to that of its impecunious neighbour.

An administrative boundary has the advantage – or disadvantage according to one's viewpoint – of inflexibility. It is particularly effective when it marks the divide between two distinct cultures, and when the areas thus defined have a strong internal organisation. Whether reflecting the past or the present, it must be taken into consideration by geographers, not *per se* but in relation to its economic and social consequences. Conversely, one might argue that many administrative boundaries are geographically deviant, and that an improved spatial understanding resulting from our research would enable us to adjust them more equitably. More systematic research into analogous situations overseas is urgently needed by France, for example, where the communes like the cantons can no longer adjust to modern economic developments and where the proposed new economic regions, in danger of being perpetuated by impending legislation, are already subject to strong criticism (Philipponeau, 1960).

Division according to dominant relationships

Many geographers have been working towards a definition of a region based on dominant relationships, but have been criticised for their hitherto unsatisfactory results. As early as 1922, Fèbvre wrote that they seemed incapable of outlining a system of geographically-based regional divisions and were content to make use of an historical framework 'like hermit crabs inside the old shells of political and administrative history'. Perroux levelled the same criticism thirty years later, accusing geographers of a lack of cooperation with economists in deciding a valid framework for the division of space.

After several meetings of the IGU Commission on Regions and Regionalisation, Dziewonski (1968) attempted to synthesise the wide range of viewpoints by suggesting three guidelines for regional classification: the region as an instrument of research, that is, as a convenient statistical unit; the region as an instrument for action, as a space subject to planning decisions; the region as both the object and the objective of geographical research. Although this classification is a purely formal one, and the first definition raises considerable problems (Bobek, 1967), it is valuable in that the third category – continually referred to during conferences in Prague, Warsaw, Liège, Strasbourg – is precisely the one with which we have so far been concerned. Bobek had already raised the issue as to whether in fact there exist spatial complexes which function as closed systems, and therefore permit a single method of regional division such that the coincidence of diverse features produces a distinctive 'geographical personality' (Vidal de la Blache). He concluded that such perfect coincidence is to be found only in complexes of the lowest order, where the basic elements are relief, soils, cultivated areas, rivers, cyclones, . . . 'in areas of a higher order there is weaker integration between the individual elements . . . we can modify a single element without radically altering the basic characteristics of the whole complex'.

Bobek was too pessimistic and his conclusion inaccurate since he examined only the degree of coincidence, whereas I would select from a whole series of coincidences those which seem fundamental, where features are so closely linked that interaction would be a more appropriate term. We should therefore identify a functional complex allowing us to define an area where

1. Interrelationships between the dominant features clearly distinguish them from features of neighbouring areas and
2. Any fundamental modification of key elements alters the appearance of the whole.

What forms of human activity should be considered in relation to this basic complex? They include development of the potential wealth of an area (mineral resources, soils as they relate to a particular agricultural system, the catchment area of a river basin such as the Tennessee Valley in the United States, . . .); the reclamation or colonisation of certain natural environments (marshlands, forest clearance, . . .); the application of two types of 'colonial agriculture' (plantations, irrigation); the planning of a transport network; the creation of growth poles of various types and functions; the circulation of commodities, capital, investment, . . .

All these activities depend to some extent on the nature of the society itself (particularly on its ethnic origin and demographic structure), on the efficiency of its enterprises (related primarily to the level of technology), and on the general orientation of its economic policies. These three characteristics, each relevant to the division of space, are themselves broadly interdependent and do not operate on a virgin environment with a once and for all impact, but as part of a continuous process of evolution.

The groups of elements will not have the same composition in countries with differing levels of economic development, and both their limits, their internal structure and the degree of cohesion will vary across time and space. With few exceptions, the limits are simply more or less clearly defined and extensive transition zones, and the geographer's task of distinguishing significant spatial divisions is rarely a simple one. Analysis of the flexibility of these zones under certain conditions, however, may provide invaluable information for planners seeking to understand the spontaneous process of regionalisation.

The traditional approach of direct observation and of comparative and integrated cartography is obviously a useful stage in this type of research but we must also utilise the whole range of statistical procedures – to which a short guide and substantial bibliography may be found in Spence and Taylor (1970).

5

Conclusion—An outline typology for the division of space

The nature of the physical environment and the impact of human activity are so varied that their potential range of interaction is almost infinite – resulting, some would claim, in a series of unique spatial complexes, the analysis of which is the geographer's major task. We have similarly stressed that the objective of geography is analysis and interpretation of the fundamental relationships dominating a specific area and distinguishing it from surrounding areas.

We should now try to draw together the numerous scattered observations and conclusions of the preceding chapters.

In areas of the world where the physical environment is so hostile that it has seldom been penetrated by man, regional divisions are based solely on physical criteria: they comprise either broad climatic zones within which the relief introduces an element of variation, or extensive mountainous or lowland areas where local variations in site, altitude and exposure in their turn create smaller but distinctive environments. In both cases the study of the basic complex is related only to associations of physical elements of which one of the most useful indices, wherever it exists, is the natural vegetation – as in the 'empty' intertropical zones of Africa or America.

Once man appears on the scene, two types of division are common. The first occurs where broad zones are determined by the physical environment and human intervention is rudimentary, due to the lack of organisation among the population or its scattered distribution. Secondly, there are zones dominated by human activity within which are found both static and functional units. The former include a favourable environment giving rise to nucleated settlement, a particular ethnic group with a high density of population and the ability to develop the land it occupies, the localised existence of supplementary resources – in fisheries, mines, a specialised technology or agricultural system, 'colonisation' by a new form of cultivation such as plantations and irrigation. Functional areas develop round a pole of economic activity (port, mining town, manufacturing centre) or an administrative or commercial town which owes its existence to a well-planned communications network. Regional development is therefore a gradual process: between the clearly-defined fragments of space is a loose fabric on which differential factors operate progressively as human societies themselves evolve. Population density is therefore a good index of local variation within this 'human' division of space (Meynen and

Hammerschmidt, 1967). With certain modifications, examples are found in both Africa and Latin America, and in certain areas of Australia and Asiatic Russia (Fig. 5.1, Type A).

A third major division is found in the highly-developed areas, where man's presence has had a profound spatial impact, either because of a large total population and high level of technology (as in western Europe and Japan) or of industrial superiority (as in the United States). Where the physical environment is on a grand scale, the major divisions are themselves vast (such as the 'belts' of the USA): where the natural features are fragmented, 'pays' such as those of the Paris Basin begin to emerge. In both cases, however, functional nodes are superimposed on the static divisions and become their fundamental characteristics – firstly because their extent and influence make them true 'crossroads', reuniting diverse fragments of the spatial mosaic (such as Lyon, linking parts of the Massif Central, the Jura, the Alpes du Nord and the lowland corridors of the Rhône–Saône, or St Louis and Chicago, on which converge the varied products of agricultural areas remarkable both for their productivity and vast acreage), and secondly, because they are supported by a whole hierarchy of intermediate centres. In a perfect system, not one but a whole series of functional divisions is superimposed on a static unit, permitting choice, competition, internal change and ultimately the provision of the highest order of central functions. Within this third category, it is the diversity of functions and the dimensions of the centre of attraction – most commonly the urban network – which constitute the external expression of the various structural systems: Fig. 5.1, Type B (Berry and Pred, 1961; Berry, 1967; Kluszka, 1969; Bobek, 1970).

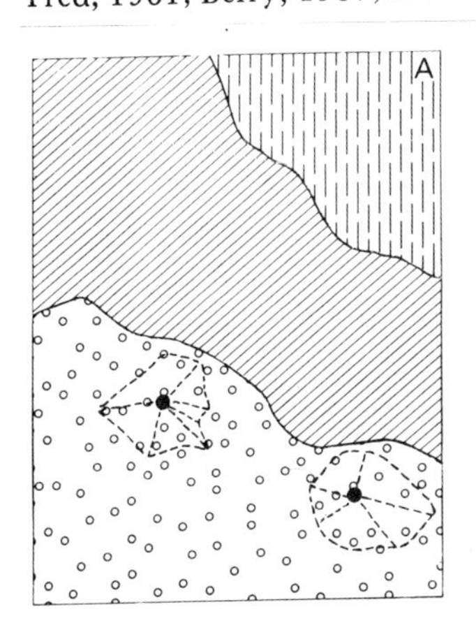

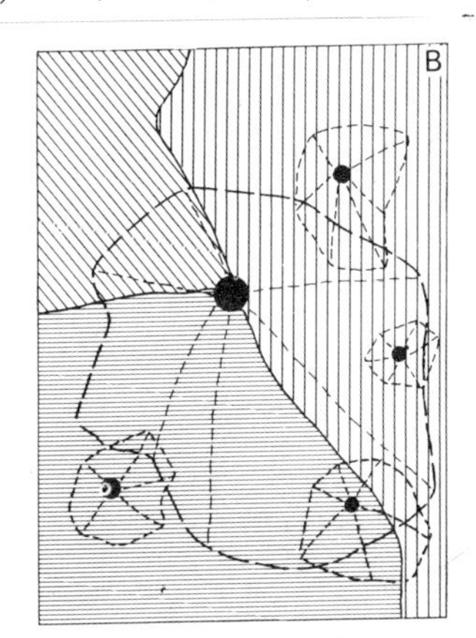

Fig. 5.1 Types of spatial division

The stippled areas are static units defined according to environmental conditions or the system of cultivation.

The circles represent towns and the dotted lines the limits of their effective influence.

Type A – area with sporadic functional relationships

Type B – area with a complex superimposition of functional on static units

Finally, there exists a fourth type of division, a geographical division par excellence in which all the characteristics coincide. The physical environment provides a precise and clearly-defined framework, the life and economy of the inhabitants have adapted to this environment, a major town has developed and is the market and nerve centre of the whole complex. The Roussillon presents an almost perfect example of such exceptional convergence.

The above would be no more than a statement of the obvious were it not for the methodological implications of the classification. We must no longer simply describe examples chosen to fit this general typology, but translate it into systematic and scientific expressions.

Parallel with a more or less subjective literary description of an area, the use of precise numerical criteria as often as possible provides a basis for comparative study. We must therefore derive measurable indices to identify significant relationships (Hagood, 1943; Berry, 1964). The field is a broad one, explored already by a number of geographers overseas and now being investigated in France, particularly in relation to the definition of spatial homogeneity and to the measurement of relationships between flows and the function of centrality. We must also derive a classification of regional types and determine the laws of evolution and development which transform them.

These are not mere academic abstractions, but indispensable steps towards both the scientific development of our discipline and any rational resource management. An understanding of the mechanisms linking particular types of environment and human intervention, analysis of the forces of attraction and development potential of an urban centre, and evaluation of the efficiency of a flow network are both rewarding to the individual research worker and provide invaluable information to planners. In place of the random establishment of new projects, a policy of *ad hoc* planning with which we are only too familiar, quantitative analysis of the environment will enable the geographer to participate in 'the discovery of spatial order' (Thompson, 1966), and to act as a consultant on all spatial aspects of development.

Bibliography

Ackerman, E. A. (1958) 'Geography as a fundamental research discipline', *Research Paper*, 53, University of Chicago, Department of Geography.
Ackoff, R. L., Gupta, S. K. and Minas, J. S. (1962) *Scientific Method: optimising applied research decisions*, New York.
Anuchin, V. A. (1960) *Teoreticheskiye Problemy Geografii*, Moscow.
Applebaum, W. (1968) *Store Location Strategy Cases*, Reading, Massachusetts.
Applebaum, W. *et al.* (1968) *Guide to Store Location Research with Emphasis on Supermarkets*, Reading, Massachusetts.
Azoulay, E. and Fried, J. (1968) *Mathématiques Préparatoires aux Sciences Economiques*, SEDES, Paris.
Barbier, B. (1969) *Villes et Centres des Alpes du Sud: étude du réseau urbain*, Gap, Hautes-Alpes.
Barbut, M. (1968) *Mathématiques des Sciences Humaines*, 2 vols., Paris.
Bauchet, P. (1955) *Les Tableaux Economiques: analyse de la région Lorraine*, Paris.
Baulig, H. (1959) 'Géographie générale et géographie régionale' in *Mélanges Géographiques Canadiens offerts à Raoul Blanchard*, Cahiers de Géographie de Québec, 6.
Beaujeu-Garnier, J. (1958) *Géographie de la Population*, Paris.
Beaujeu-Garnier, J. (1969) *Trois Milliards d'Hommes*, 2nd edn., Paris.
Beaujeu-Garnier, J. (1973) 'Essai sur l'action humaine' in *Etudes de Géographie Tropicale offertes à Pierre Gourou*, Paris–Hague.
Beaujeu-Garnier, J. and Andan, O. (1971) 'Transports et régionalisation: essai méthodologique' in *Bulletin de la Section de Géographie du Ministère de l'Education Nationale*, Paris.
Berry, B. J. L. (1958) 'A note concerning methods of classification', *Annals of the Association of American Geographers*, 48.
Berry, B. J. L. and Garrison, W. L. (1958) 'The functional bases of the central place hierarchy', *Economic Geography*, 34.
Berry, B. J. L. and Pred, A. (1961) *Central Place Studies: a bibliography of theory and applications*, Regional Science Research Institute, Philadelphia.
Berry, B. J. L. (1964) 'Approaches to regional analysis: a synthesis', *Annals of the Association of American Geographers*, 54.
Berry, B. J. L. (1966) 'Essays on commodity flows and the spatial structure of the Indian economy', *Research Paper*, 111, University of Chicago, Department of Geography.

Berry, B. J. L. (1967) *Geography of Market Centers and Retail Distribution*, Englewood Cliffs, New Jersey.

Berry, B. J. L. (1968) 'Approaches to regional analysis: a synthesis' in Berry and Marble (1968).

Berry, B. J. L. and Marble, D. F. (1968) *Spatial Analysis*, Englewood Cliffs, New Jersey.

Birot, P. (1950) *Le Portugal*, Paris.

Birot, P. (1965) *Formations Végétales du Globe*, SEDES, Paris.

Blanchard, R. (1938–56) *Les Alpes Occidentales*, 7 vols., Grenoble.

Bobek, H. (1967) 'Some remarks on basic concepts in economic regionalisation' in *Economic Regionalisation*, Publishing House of the Czechoslovak Academy of Sciences, Prague.

Bobek, H. (1967) 'The hierarchy of central places and their hinterlands in Austria and their role in economic regionalisation', *Economic Regionalisation*, op. cit.

Bobek, H. (1970) 'Ausgliederung der Strukturgebiete der Osterreichischen Wirtschaft' in *Strukturanalyse des Osterreichischen Bundesgebietes*, 3, Osterreichischen Gesellschaft für Raumforschung und Raumplannung, Vienna.

Bobek, H. (1970) 'Die zentralen Orte und ihre Versorgungsbereiche' in ibid.

Bonetti, E. (1955) 'Espaces économiques et paysages économiques', *Monografie*, 14, Collana di Monografie dell'Instituto di Geografia dell' Universita di Trieste.

Bonnamour, J. (1965) *Le Morvan. La Terre et les Hommes: essai de géographie agricole*, Paris.

Boudeville, J-R. (1957) 'Contribution à l'étude des pôles de croissance brésiliens', *Cahiers de l'Institut de Science Economique Appliquée*, Série F, 10.

Boudeville, J-R. (1961) *Les Espaces Economiques*, Paris.

Boudeville, J-R. (1968) *L'Espace et les Pôles de Croissance*, Paris.

Böventer, E. von (1969) 'Walter Christaller's central places and peripheral areas', *Journal of Regional Science*, 9.

Braithwaite, R. B. (1960) *Scientific Explanation*, New York.

Brookfield, H. (1964) 'Questions on the human frontiers of geography', *Economic Geography*, 40.

Brunet, R. (1968) 'Les phénomènes de discontinuité en géographie', *Mémoires et Documents*, nouvelle série, 7, CNRS, Paris.

Brunet, R. (1969) 'L'étude des quartiers ruraux', *Revue de Géographie des Pyrénés et du Sud-Ouest*, 40.

Brunhes, J. (1956) *La Géographie Humaine*, abridged edition, Paris.

Brush, J. E. (1953) 'The hierarchy of central places in southwestern Wisconsin', *Geographical Review*, 43.

Bruyelle, P. (1970) 'L'influence urbaine en milieu rural dans la région du Nord' in *Commerce et Services*, Paris.

Bunge, W. (1966) 'Theoretical Geography', *Lund Studies in Geography*, Series C, 1.

Burton, I. (1968) 'The quantitative revolution and theoretical geography' in *Spatial Analysis*, eds B. J. L. Berry and D. F. Marble, Englewood Cliffs, New Jersey.

Chabot, G., Clozier, R. and Beaujeu-Garnier, J. (1957) *Géographie Française au Milieu du XXe Siècle*, Paris.
Champier, L. (1956) 'La première colonisation de la montagne jurassienne' *Revue de Géographie de Lyon,* 31.
Chisholm, M. D. I. (1960) 'The geography of commuting', *Annals of the Association of American Geographers,* 50.
Chisholm, M. D. I. (1967) 'General systems theory and geography', *Transactions, Institute of British Geographers,* 42.
Cholley, A. (1939–40) 'Régions naturelles et régions humaines', *Information Géographique,* 2.
Cholley, A. (1948) 'Points de vue géographiques', *Information Géographique,* 3 and 4.
Chorley, R. J. and Haggett, P. (1965) *Frontiers in Geographical Teaching*, London.
Chorley, R. J. and Haggett, P. (1967) *Models in Geography*, London.
Christaller, W. (1933) *Die Zentralen Orte in Süddeutschland*, Jena.
Christaller, W. (1966) *Central Places in Southern Germany*, English translation, Englewood Cliffs, New Jersey.
Church, R. J. Harrison (1969) 'The Firestone rubber plantation in Liberia', *Geography,* 54.
Claval, P. (1964) 'Essai sur l'évolution de la géographie humaine', *Cahiers de Géographie de Besançon,* 12.
Claval, P. (1966) 'Géographie et économie', *Cahiers de sociologie économique,* 15, Le Havre.
Claval, P. (1968) *Régions, Nations, Grands Espaces: géographie générale des ensembles territoriaux*, Paris.
Cole, J. P. and King, C. A. M. (1968) *Quantitative Geography*, London.
Cotten, A. M. (1972) 'Le réseau urbain en Côte d'Ivoire' in *La Croissance Urbaine en Afrique Noire et à Madagascar*, CNRS, Paris.
Curry, L. (1967) 'Central places in the random spatial economy', *Journal of Regional Science,* 7, supplement.
Dacey, M. F. (1962) 'Analysis of central place and point patterns by a nearest neighbour method' in *IGU Symposium in Urban Geography*, ed. K. Norborg, Lund.
Dacey, M. F. (1965) 'The geometry of central place theory', *Geografiska Annaler*, Series B, 47.
Dacey, M. F. (1966) 'A probability model for central place location', *Annals of the Association of American Geographers,* 56.
Darby, H. C. (1962) 'The problem of geographical description', *Transactions, Institute of British Geographers,* 30.
Davis, W. M. (1899) 'The geographical cycle', *Geographical Journal*, 14.
Deffontaines, P. (1953) 'Le rang: type de peuplement rural au Canada Français', *Cahiers de Géographie de l'Université de Laval*, Québec, 2.
Derruau, M. (1949) *La Grande Limagne Auvergnate et Bourbonnaise*, Clermont-Ferrand.
Despois, J. (1965) 'La géographie régionale en Algérie' in *Atti del Primero Congresso Internationale di Studi Nord-Africani*, Cagliari.
Dézert, B. (1969) *La Croissance Industrielle et Urbaine de la Porte d'Alsace*, SEDES, Paris.

Dickinson, R. E. (1952) *City, Region and Regionalism*, 2nd edn., London.
Dickinson, R. E. (1957) 'The geography of commuting: the Netherlands and Belgium', *Geographical Review*, **47**.
Dickinson, R. E. (1967) *The City Region in Western Europe*, London.
Dollfus, O. (1970) *L'Espace Géographique*, Paris.
Downs, R. M. (1970) 'Geographic space perception', *Progress in Geography*, 2.
Dugrand, R. (1963) *Villes et Campagnes du Bas-Languedoc*, Paris.
Durand-Dastes, F. (1965) *Géographie de l'Inde*, Paris.
Durand-Dastes, F., Cribier, F. and Drain, M. (1967) *Initiation aux Exercices de Géographie Régionale*, SEDES, Paris.
Dziewonski, K. (1968) 'Economic Regionalisation', *Geographia Polonica*, **15**.
Economic Regionalisation (1965) Proceedings of the 4th General Meeting of the Commission on Methods of Economic Regionalisation of the International Geographical Union, Prague.
Fèbvre, L. (1922) *La Terre et l'Evolution Humaine*, Paris.
Finch, V. C. and Trewartha, G. T. (1936) *Elements of Geography*, New York.
Flatres, P. (1957) *Géographie Rurale des Quatre Contrées Celtiques: Irlande, Galles, Cornwall et Man*, Rennes.
Forsberg, F. R. (1962) *The Island Ecosystem*, Honolulu.
Freeman, T. W. (1961) *A Hundred Years of Geography*, London.
Freeman, T. W. (1966) *The Geographer's Craft*, Manchester.
Gadille, R. (1967) *Le Vignoble de la Côte Bourgignonne: fondements physiques et humains d'une viticulture de haute qualité*, Paris.
Gallois, L. (1908) *Régions Naturelles et Noms de Pays*, Paris.
Gendarme, R. (1954) *La Région du Nord: essai d'analyse économique*, Paris.
George, P. (1963) 'L'aspect géographique de la division régionale' in *La Région*, rapport ronéoté du Colloque de Lyon, CNRS, Lyon.
George, P. (1968) *L'Action Humaine*, Paris.
Gilbert, E. W. (1960) 'The idea of the region', *Geography*, **45**.
Gluckman, M. (1964) *Closed Systems and Open Minds: the limits of naïvety in social anthropology*, Chicago.
Golledge, R. and Amedeo, D. (1968) 'On laws in geography', *Annals of the Association of American Geographers*, **58**.
Gottmann, J. (1952) 'L'aménagement de l'espace: planification régionale et géographie', *Cahiers de la Fondation Nationale des Sciences Politiques*, Paris.
Gottmann, J. (1966) *Essais sur l'Aménagement de l'Espace Habité*, Paris–Hague.
Gottmann, J. (1967) *Megalopolis*, Cambridge, Massachusetts.
Gould, P. (1970) 'Computers and spatial analysis: extensions of geographic research', *Geoforum*, 1.
Gourou, P. (1970) *L'Afrique*, Paris.
Gower, J. C. (1967) 'Multivariate analysis and multidimensional geometry', *The Statistician*, 17.
Gregory, S. (1968) *Statistical Methods and the Geographer*, 2nd edn., London.

Gribaudi, D. (1965) 'Tendenze coesive nei più recenti sviluppi della geografia Italiana', *Rivista Geografia Italiana,* 72.
Grigg, D. B. (1965) 'The logic of regional systems', *Annals of the Association of American Geographers,* 55.
Grigg, D. B. (1967) 'Regions, models and classes' in *Models in Geography*, eds R. J. Chorley and P. Haggett, London.
Grigoriev, A. A. (1954) 'Geograficheskaya zonalinosti i nekotorie ee zakonomernosti', *Izvestiya Akademii Nauk SSSR*, Seriya geograficheskaya, 6.
Grigoriev, A. A. (1956) 'Sur l'état contemporain de la théorie de la zonalité dans la nature' in *Essais de Géographie*, Academy of Sciences of the USSR, Moscow.
Guichard, O. (1965) *Aménager la France*, Paris–Geneva.
Guitton, H. (1955) Introduction à la thèse de Ponsard, *Economie et Espace*, SEDES, Paris.
Hägerstrand, T. (1967) 'The computer and the geographer', *Transactions, Institute of British Geographers,* 42.
Hägerstrand, T. (1969) 'Geographic measurements of migration' in *Human Displacements: measurements, methodological aspects*, ed. J. Sutter, Monaco.
Haggett, P. (1965) *Locational Analysis in Human Geography*, London.
Hagood, M. J. (1943) 'Statistical methods for the delineation of regions applied to data on agriculture and population', *Social Forces,* 21.
Harris, C. D. (1954) 'The market as a factor in the localisation of industry in the United States', *Annals of the Association of American Geographers,* 44.
Hartshorne, R. (1939) 'The nature of geography', *Annals of the Association of American Geographers,* 29.
Hartshorne, R. (1959) *Perspective on the Nature of Geography*, Monograph Series of the Association of American Geographers, Chicago.
Harvey, D. (1969) *Explanation in Geography*, London.
Hettner, A. (1919) 'Die Einheit der Geographie in Wissenschaft und Unterricht', *Geographische Abende,* 1.
Huntington, E. (1927) *The Human Habitat*, New York.
Isard, W. (1956) *Location and Space Economy: a general theory relating to industrial location, market areas, land use, trade and urban structures*, Cambridge, Massachusetts.
Isard, W. (1960) *Methods of Regional Analysis*, New York.
James, P. E. and Jones, C. F. (1954) *American Geography: inventory and prospect*, Syracuse.
Jensen, M. (1951) *Regionalism in America*, Madison, Wisconsin.
Jensen, R. G. and Karaska, G. J. (1969) 'The mathematical thrust in geography', *Journal of Regional Science,* 9.
Juillard, E. (1962) 'La région: essai de définition', *Annales de Géographie,* 71.
Kao, R. C. (1968) 'The use of computers in the processing and analysis of geographical information' in *Spatial Analysis*, eds B. J. L. Berry and D. F. Marble, Englewood Cliffs, New Jersey.
Kayser, B. (1966) *Types de Régions en Pays en cours de Développement*,

Communication à la Conférence Internationale de Géographie de Mexico, Mexico City.
Kayser, B. (1967) *Les Transformations de la Structure Régionale par l'Economie Commerciale dans les Pays Sous-Développés*, Communication au Colloque de Strasbourg, CNRS, Strasbourg.
Kendall, M. G. (1961) 'Natural law in the social sciences', *Journal of the Royal Statistical Society*, Series A, 124.
King, L. J. (1961) 'A multivariate analysis of the spacing of urban settlements in the United States', *Annals of the Association of American Geographers*, 51.
Klatzmann, J. (1956) *La Localisation des Cultures et des Productions Animales en France*, Paris.
Kluczka, G. (1969) 'Zentrale Orte und ihre Einzugsbereiche in der Bundesrepublik Deutschland', *Berichte zur Deutschen Landeskunde*, 42.
Kolotievskij, A. M. (1967) *Voprosy Teorii i Metodiki Ekonomiceskogo Rajonirovanija*, Riga.
Korach, M. (1966) 'The Science of Industry' in *The Science of Science*, eds M. Goldsmith and A. L. Mackay, London.
Labasse, J. and Laferrère, M. (1960) *La Région Lyonnaise*, Paris.
Labasse, J. (1966) *L'Organisation de l'Espace*, Paris.
Laferrère, M. (1960) *Lyon, Ville Industrielle*, Paris.
Lefèvre, M. (1965) 'Le concept de région géographique', *La Géographie*, 17.
Lehmann, E. (1967) 'Regionale Geographie und Naturräumliche Gliederung' in *Probleme der Landschaft-ökologischen Erkundung und Naturräumlichen Gliederung*, Leipzig.
Lehmann, E. (1968) 'Die Typisierung als Problem der kartographischen Darstellung im "Atlas DDR"', *Petermanns Geographische Mitteilungen*, 112.
Le Lannou, M. (1949) *La Géographie Humaine*, Paris.
Libault, A. (1970) *Communication sur l'Atlas de São Paulo*, unpublished, Madrid.
Lösch, A. (1936) *Wirtschaftsgebiete als Grundlage des Internationalen Handels*, unpublished lectures, Bonn.
Lösch, A. (1940) *Die Räumliche Ordnung der Wirtschaft*, Jena.
Lösch, A. (1954) *The Economics of Location*, English translation, New Haven, Connecticut.
Lowenthal, D. (1961) 'Geography, experience and imagination: towards a geographical epistemology', *Annals of the Association of American Geographers*, 51.
Lynch, K. (1960) *The Image of the City*, Cambridge, Massachusetts.
McDonald, J. R. (1966) 'The region: its conception, design and limitation', *Annals of the Association of American Geographers*, 56.
Mathieson, R. S. (1969) 'The Soviet contribution to regional science', *Journal of Regional Science*, 9.
May, J. M. (1958) *The Ecology of Human Disease*, New York.
Médevielle, G. (1970) *La Micro-Polarisation de l'Espace autour d'un Bourg Charentais: Champagne-Mouton*, unpublished, Paris.
Mensching, H. (1951) 'Une accumulation post-glaciaire provoquée par des défrichements', *Revue de Géomorphologie Dynamique*, 2.

Methods of Economic Regionalisation (1964) Geographia Polonica, **4**, Polish Academy of Sciences, Warsaw.
Meynen, E. and Hammerschmidt, A. (1967) 'Zur Karte der Bevölkerungdichte der Bundesrepublik Deutschland nach naturräumlichen Einheiten', *Berichte zur Deutschen Landeskunde,* **39**.
Meynier, A. (1969) *Histoire de la Pensée Géographique en France*, Paris.
Mikesell, M. W. (1969) 'The borderland of geography as a social science' in *Interdisciplinary Relationships in the Social Sciences*, eds M. Sherif and C. W. Sherif, Chicago.
Millies-Lacroix, A. (1965) 'L'instabilité des versants dans le domaine rifain', *Revue de Géomorphologie Dynamique,* **15**.
Minshull, R. (1967) *Regional Geography: theory and practice*, London.
Montefiore, A. C. and Williams, W. W. (1955) 'Determinism and possibilism', *Geographical Studies,* **2**.
Motte, B. (1960) *Compte Rendu de l'Assemblée Parlementaire Européenne*, Strasbourg, 9 May.
National Academy of Sciences (1965) *The Science of Geography*, Washington.
Nicolaï, H. (1963) *Le Kuslu*, Brussels.
Nougier, L-R. (1959) *Géographie Humaine Préhistorique*, Paris.
Odum, H. W. and Moore, H. E. (1938) *American Regionalism*, New York.
Pedelaborde, P. (1970) *Les Mathématiques Elémentaires Appliquées à la Géographie Physique*, Paris.
Pélissier, R. (1966) *Les Paysans du Sénégal: les civilisations agraires du Cayor à la Casamance*, St Yrieux.
Pélissier, R. (1970) *Les Effets de l'Opération Arachide-Mil dans les Régions de Thiès, Djourbel et Kaolack (République du Sénégal)*, mimeo.
Penck, A. (1919) 'Der Gipfelflur der Alpen', *Sitzungberichte Preussischer Akademie Wissenschaft*, **17**.
Penck, W. (1924) *Die Morphologische Analyse: ein Kapital der physikalischen Geologie*, Stuttgart.
Perroux, F. (1950) 'Economic space: theory and applications', *Quarterly Journal of Economics,* **64**.
Perroux, F. (1954) *L'Europe sans Rivages*, Paris.
Philipponeau, M. (1960) *Géographie et Action: introduction à la géographie appliquée*, Paris.
Piaget, J. (1950) *Introduction à l'Epistémologie Génétique*, 3 vols., Paris.
Piaget, J. (1968) *Logique et Connaissance Scientifique*, Paris.
Piatier, A. (1970) *Atlas d'Attraction Urbaine*, Paris.
Pinchemel, P. and Carrière, F. (1963) *Le Fait Urbain en France*, Paris.
Piveteau, J. L. (1969) 'Le sentiment d'appartenance régionale en Suisse', *Revue de Géographie Alpine*, **62**.
Poincaré, J. H. (1908) *Science et Méthode*, Paris.
Ponsard, C. (1955) *Economie et Espace*, Paris.
Ponsard, C. (1958) *Histoire des Théories Economiques Spatiales*, Paris.
Ponsard, C. (1969) *Un Modèle Topologique d'Equilibre Economique Interrégionale*, Paris.
Précheur, C. (1959) *La Lorraine Sidérurgique*, Paris.

Précheur, C. (1969) *Les Industries Françaises à l'Heure du Marché Commun*, SEDES, Paris.
Problèmes de Formation et d'aménagement du réseau urbain (1965) Compte rendu du 2e colloque de géographie franco-polonais, Warsaw.
Prothero, R. M. (1965) *Migrants and Malaria*, London.
Reynaud, A. (1970) 'Les sens du mot "géographie"', *Travaux de l'Institut de Géographie de Reims*, 3.
Quermonne, J-L. (1964) 'Planification régionale et régions administratives', *Cahiers de la Fondation Nationale des Sciences Politiques*, Paris.
Rochefort, M. (1960) *L'Organisation Urbaine de l'Alsace*, Paris.
Rochefort, M. (1970) *Rapport sur l'Organisation Urbaine en Afrique Noire et à Madagascar*, CNRS, Bordeaux.
Roscher, W, (1865) *Studien über die Naturgeste welche den zweckmässigen Standort der Industriezweige bestimmen*, Heidelberg.
Salichtchev, K. A., Saushkin, Y. G. and Guseva, I. N. (1970) *Synthetic Social-Economic Maps*, Moscow.
Sauer, C. O. (1927) 'Recent developments in cultural geography' in *Recent Developments in the Social Sciences*, ed. E. C. Hayes, Philadelphia.
Sauer, C. O. (1963) *Land and Life: a selection from the writings of Carl Ortwin Sauer*, ed. J. Leighly, Berkeley.
Saushkin, Y. G. (1965) 'The today and tomorrow of geography', *Soviet Geography*, 6.
Sautter, G. (1966) *De l'Atlantique au Fleuve Congo: une géographie du sous-peuplement*, Paris.
Sautter, G. (1967) *La Région Traditionnelle en Afrique Tropicale*, Communication au Colloque de Strasbourg, CNRS, Strasbourg.
Schmithüsen, J. (1963) *Was ist eine Landschaft*, Wiesbaden.
Schmitt, E. (1969) *Deutschland*, vol. 1.
Scott, P. (1970) *Geography and Retailing*, London.
Sestini, A. (1963) *Appunti per una Definizione del Paesagio Geografico*, Naples.
Sorre, M. (1957) *Rencontre de la Géographie et de la Sociologie*, Paris.
Sotchava, V. D. (1956) 'Les principes de la division physico-géographique des territoires' in *Essais de Géographie*, Academy of Sciences of the USSR, Moscow.
Spence, N. A. and Taylor, P. J. (1970) 'Quantitative methods in regional taxonomy', *Progress in Geography*, 2.
Stamp, L. D. (1957) 'Geographical agenda: a review of some tasks awaiting geographical attention', *Transactions, Institute of British Geographers*, 23.
Stevens, B. H. and Brackett, C. A. (1968) 'Regionalisation of Pennsylvania counties for development planning', *Geographia Polonica*, 15.
Stewart, J. Q. (1947) 'Empirical mathematical rules concerning the distribution and equilibrium of population', *Geographical Review*, 37.
Stone, R. (1960) 'A comparison of the economic structure of regions based on the concept of distance', *Journal of Regional Science*, 2.
Thibault, A. (1967) *Villes et Campagnes de l'Oise et de la Somme*, Beauvais-Amiens.
Thomas, E. N. (1962) 'The stability of distance, population, size relation-

ships for Iowa towns from 1900 to 1950' in *IGU Symposium in Urban Geography*, ed. K. Norborg, Lund.

Thomas, E. N. and Anderson, D. L. (1965) 'Additional comments on weighting values in correlation analysis of areal data', *Annals of the Association of American Geographers*, 55.

Thompson, J. H. (1966) *Geography of New York State*, Syracuse.

Toussaint, M. (1951–52) 'Le territoire et les limites de la Civitas Verodunensium', *Bulletin Archaéologique*.

Tricart, J. and Cailleux, A. (1950) *Le Modelé des Régions Périglaciaires, Traité de Géomorphologie*, 2, Paris.

Tricart, J. (1953) 'Géomorphologie dynamique de la steppe russe', *Revue de Géomorphologie Dynamique*, 4.

Tricart, J. (1953) 'Erosion naturelle et érosion anthropogène à Madagascar', *Revue de Géomorphologie Dynamique*, 4.

Tricart, J. and Cailleux, A. (1957) *The Continental Plates, Traité de Géomorphologie*, 12, Paris.

Tricart, J. (1968) *Précis de Géomorphologie: géomorphologie structurale*, Paris. (English translation, *Structural Geomorphology*, London, 1974.)

Uhlig, H. (1970) 'Organisations-Plan und System der Geographie', *Geoforum*, 1.

Ullman, E. L. (1957) *American Commodity Flow*, Seattle.

Vasilevskii, V. (1966) 'Mathematical models in economic geography', *Sovetskaya Entsiklopediia*, 5, Moscow.

Vennetier, P. (1965) 'Les hommes et leurs activités dans le nord du Congo-Brazzaville', *Cahiers ORSTOM*, 2.

Von Thünen, J. H. (1826–63) *Der isolierte Staat in Beziehung auf Landwirtschaft und Nationalökonomie*, 3 vols., Rostock.

Weber, A. (1909) *Uber den Standort der Industrie*, Tübingen.

Weber, M. (1949) *The Methodology of the Social Sciences*, Glencoe, Illinois.

Whittesley, D. (1954) 'The regional concept and the regional method' in *American Geography: inventory and prospect*, ed. P. E. James and G. F. Jones, Syracuse.

Winkler, E. (1970) 'A possible classification of the geosciences', *Geoforum*, 1.

Zobler, L. (1957) 'Statistical testing of regional boundaries', *Annals of the Association of American Geographers*, 47.

Index

Aerial views, 63, 64, 78
Afforestation, 14–15
Allier valley, 60–1
Alps, 10–11, 92, 94–5
Anthropology, 3
Anuchin, V. A., 4, 8, 38
Appalachians, 83, 93
Applebaum, William, 16, 17
Association of American Geographers, 82

Baranski, Professor, 37
Bauchet, P., 58
Berkeley school, 47
Berry, B. J. L., 77
Blanchard, Raoul, 10
Bobek, H., 102–3
Bomer, B., 29
Bonnamour, Jacqueline, 13
Boudeville, J-R., 59
Böventer, E. von, 23
Braithwaite, R. B., 34
Brazil, 14, 15, 31, 61
Brookfield, H., 23, 47
Brunet, R., 31, 61, 70, 90
Brunhes, J., 71, 84
Bunge, W., 25, 36, 37

Canada, 17
Cartography, 2, 41, 63
Case studies, 23, 24, 34–5
Causality, 30, 33
Central place theory, 23–4
Champagne-Mouton, 65
Cholley, A., 20, 58, 70
Chorley, R. J., 34
Christaller, W., 4, 23, 36, 52, 54, 55
Claval, P., 86–7, 101
Climate, 91–3
Climatic change, 14
Climatology, 2, 3, 7, 8, 20, 68

Coal mining, 50
Comparative method, 25
Complementarity, 90, 94
Computers, 27, 41–2
Conceptualisation, 24–5
Crop rotation, 15
Cuestas, 17, 28

Davis, W. M., 26, 29, 37
Deductive method, 23, 24
Deffontaines, P., 98
Deforestation, 15
Demangeon, A., 29, 67
De Martonne, E., 20–1, 67
Derruau, M., 61
Descriptive method, 25
Despois, J., 81
Dézert, B., 57
Division of space, 51, 86, Ch. 4
  administrative, 100–2
  based on dominant relationships, 102–3
  and human activity, 97–9, 104
  physiographic, 91–7, 104
  typology of, 104–6
  visual, 99–100
Dunkirk, 11, 50
Dziewonski, K., 102

Economic geography, 3–4, 5, 45, 52, 54, 72
Economic space, 52–60
  definitions of, 59
  regions, 54–8
Economic theory, 44
  and geography, 2–5 *passim*, 52–60
  and spatial concepts, 52–60
  time dimension in, 31–2
Ecosystem, 33
England, geography in, 37

Environment
  and division of space, 91, 96–7
  and the firm, 58–9
  and human activity, 9–16, 33, 97–9, 103, 104
Erosion, 14, 29
  normal cycle of, 23, 29
European Economic Community, 57
Explanation, techniques of, 30–3

Factor analysis, 41
Fèbvre, L., 102
Flatres, P., 32
France, geography in, 3, 36, 55
Freeman, E. A., 5
Freeman, T. W., 7, 42
Functional systems, 65, 72–9

Gadille, Rolande, 12
Gallois, L., 84
Gendarme, R., 55–7
Geography; *see also* Economic, Human, Physical, Regional, Urban geography
  applied, 6
  criticism of, 1–2
  diversification of, 67–8
  general, 67–9, 70, 72
  and mathematics, 35–41
  meaning of, 4–5
  role and scope of, 2, 5–6, 16–18, 19–22, 44–5
  as science, 42–3, 49
  social, 5
  specialisation in, 2–3, 6–7, 14, 21–2, 68
  unity of, 4–9, 10, 19, 67
Geology, 2, 3, 5, 7, 21
  in petroleum exploitation, 17
Geometry, 41
Geomorphology, 2, 5–7 *passim*, 14, 15, 20, 68, 89
  in petroleum exploitation, 16–17
George, P., 84
Gerasimov, S., 4
Germany, 15
  location theory in, 53
Gilbert, E. W., 20, 72, 82
Glacial valleys, 17
Gluckman, M., 21
Gottmann, J., 59, 84
Gould, P., 41
Gourou, P., 91
Gribaudi, D., 6, 81
Growth pole, 78, 88, 103
Guanabara, 11

Haggett, P., 34
Hartshorne, R., 5, 8, 36, 46, 67
Harvey, D., 8, 24, 34, 39
Hettner, A., 71
History, 3, 5, 8
  influence of, 100–1
Homogeneity, 60–3, 66, 76, 90, 106
  absolute and relative, 61–2
  recurrent, 62–3
Hoover, 4
Human activity, *see* Environment
Human geography, 2, 3, 6, 7, 20
  and physical environment, 9–16
Humboldt, Alexander von, 4
Huntington, E., 9, 30
Huxley, Julian, 21
Hydrology, 14

IFOP survey, 80
IGU
  Commission on Regions and Regionalisation, 102
  Congress 1938, 47
Ile-de-France, 65
Inductive method, 23–4
Industrial Revolution, 11, 55
Industries, siting of, 11–12, 31; *see also* Location theory
INSEE, 55
Inselbergs, 17
Interdisciplinary research, 21–2
Iran, 12
Iron ore, 50
Isard, W., 4, 52, 55

James, Preston, 47, 82, 83
Japan, 92
Juillard, E., 47

Kansas, 17
Kao, R. C., 41
Karst landscapes, 20, 67
Kayser, B., 84
Kinetic regions, 82
Kolotievskij, A. M., 26

Labasse, J., 80, 87
Lancashire, 11–12, 82
Land Management, 16
Landscape, 46–8
Lefèvre, M., 84
Le Lannou, M., 6, 10, 83, 97
Leningrad school, 4
Limousin, 29–30, 65
Location theory, 53–4, 58–9
Loire valley, 60–1

Lorraine, 12, 50, 58, 70, 101
Lösch, A., 4, 54–5
Lyon, 11, 33

Maps, 46
Market areas, 53–5, 60
Massif Central, 13, 60–1
Mathematical geography, 40–1
Mathematics, *see* Quantitative methods
Methodology, 7–9, 16, 22, Ch. 2, 55
  explanatory, *see* Explanation, techniques of
  general processes of, 23–6
  models, *see* Models
  observation, *see* Observation
Migration, 50–1
Models, 23, 25, 33–5, 38, 55
  conclusive, 35
  explanatory, 35
  exploratory, 48
  reference, 34
Monographs, 2, 20–1, 25, 26–7, 49, 64
Morvan, 13, 80
Moscow school, 4

New England, 31
Network analysis, 41
Northumberland, 11
Nougier, L-R., 32

Observation, 26–30
Oil fields, 16–17

Palander, 53
Paris, 41, 78–9
Paris basin, 96, 97
Pedology, 13
Periglacial processes, 48–9
Perroux, F., 52, 58–60, 76, 102
Photographs, 46
Physical geography, 2, 6, 8, 26, 72
  experimental work in, 27
  and human activity, 9–16
Physiography, 3
Piaget, J., 7, 43
Piatier, A., 55
Pinchemel, P., 36
Piveteau, J. L., 80
Polarisation, 59, 65, 76, 84, 101
Pollution, 12
Ponsard, C., 58, 59
Population geography, 20, 42, 68, 89
Précheur, C., 70
Pred, A., 21, 24
Pre-history, 32
Probability theory, 30

Quantitative geography, 38–9, 40
Quantitative methods, 35–42, 49, 56, 57
Quantitative revolution, 35, 36, 42

Reconcavo de Bahia, 84–6
Region, 90
  definitions of, 79–81, 82, 97, 102
  economic, 54–8, 87–8
  historical, 101–2
  and regionalisation, 86–8
  types of, 81–6
Regional balance, 58
Regional geography, 10, 51, 56–8, 65–6, 69–72
Regional Science Association, 4
Relief, 91–4
Ritter, Karl, 9
Rochefort, M., 84
Rodoman, 83
Rognon, Pierre, 16, 19
Roscher, W., 53
Roubaix–Tourcoing, 12, 31

Sahara, 16, 17
Salitchev, K. A., 41
Sandstone, 17
Saône lowlands, 94, 96
Sauer, C. O., 43, 46
Saushkin, Y. G., 4
Sautter, G., 97
Scale, 64, 93–4
Schmithüsen, J., 47
Sestini, A., 47
Simulation models, 27
Sociology, 2, 3, 5, 21, 30, 44
Soil erosion, *see* Erosion
Sorre, M., 81
Sotchava, V. D., 83
Space, Ch. 3; *see also* Division of space; Homogeneity; Region; Space elements
  analysis of, 4, 31, 63–79
  coherent, 48–50
  complex, 44–5
  concrete, 45–8
  economic, *see* Economic space
  and functional linkages, 72–9, 104–5
  static, 60–3, 66, 72–9, 104–5
  variable and changing, 50–1
Space elements
  comparison of, 67–9
  grouping of, 69–72, 89–90
  identification of, 63–6
  static and functional, 72–9, 104–5
Spence, N. A., 41

Statistical geography, 39–40, 49
Stewart, J. Q., 36
Supermarkets, 16, 17–18

Taylor, P. J., 41
Technology, 50
Terminology, 22, 28–9
Theory, 29–30
  and models, 34
Topology, 41
Transport geography, 20
Tricart, J., 8, 15

United States, 16, 61–2
  colonisation of, 93
  geography in, 2–3, 36, 52
University of Chicago, 3
Urban environments, 99
Urban geography, 20, 21, 68–9, 70
USSR, 14–15
  geography in, 3–4, 37–8, 52, 83

Vasilevskii, V., 38
Vidal de la Blache, P., 30, 47, 56, 71
Viticulture, 12–13
Von Thünen, J. H., 4, 53, 55

Water, 12, 93
Weber, Max, 4, 24, 50, 53, 55
Winkler, E., 7